Triumph Motorcycle Owner's Handbook

By

A. St. J. Masters

Originally published in 1951

ISBN: 978-1-940001-49-4

Sportsman's Vintage Press
2015

TRIUMPH MOTORCYCLES

A PRACTICAL GUIDE COVERING
ALL MODELS FROM 1937

By
A. St. J. MASTERS
Service Manager
Triumph Engineering Co., Ltd.

WITH OVER 100 ILLUSTRATIONS

Triumph Trophy - 498 c.c. OHV Vertical Twin

PREFACE

In compiling this manual the author has endeavoured to provide full maintenance information regarding Triumph motor cycles manufactured from 1937.

The manual will be of considerable assistance to garages and workshops where Triumph machines are repaired and also to the private owner who carries out his own maintenance.

Many motor cyclists are efficient fitters, but a word of warning is necessary to those whose experience is limited. This book gives only maintenance information and is not a motor cycle repair manual instructing the rider in fitting pure and simple. The inexperienced owner should avoid those jobs which require a high degree of skill or he may find himself, as others have done, with a finished job which is mechanically inferior to the assembly on which he started work. Such owners will frequently be able to find a friend who has the right kind of knowledge, and many motor cycle clubs are able to provide skilled assistance.

The manual should be used in conjunction with the spare part list published for the particular model, copies of which are available from the manufacturers.

The author would like to express his appreciation to the Triumph Engineering Company Limited, Messrs. Joseph Lucas Limited, The British Thomson-Houston Company Limited and Messrs. Amal Limited for the valuable assistance they have given him.

A. St. J. M.

NOTE ON THE THIRD EDITION

The opportunity has been taken to include necessary information on the latest models.

CONTENTS

CONTENTS

CONTENTS

CHAPTER I

SINGLE-CYLINDER ENGINES

TRIUMPH single-cylinder engines manufactured since 1936 are of the 250-, 350- and 500-c.c. capacity O.H.V., and 350-, 500- and 600-c.c. side-valve type. They are of the same basic design with a similar crankcase layout, the crankshaft running on three ball bearings with double row, uncaged roller big-ends, dry-sump lubrication and ignition from a magdyno. The O.H.V. models are fitted with detachable aluminium rocker-boxes, with the exception of the Army model 3HW where the rocker-boxes are cast integral with the cylinder-head. The engines are designed to require no special tools in order to service them, and it will be found that everything normally required will be available in the well-equipped workshop or the tool-kit of the machine.

Lubrication

The single-cylinder models are lubricated on the dry-sump system. Oil is fed to the oil-pump, pressure side, from the oil-tank by gravity and suction. To reach the oil-pump, it has to pass through a filter which is fitted to the large hexagonal nut securing the oil-pipe union to the tank. After leaving the pump, the oil passes to the timing side of the crankshaft assembly. The assembly is suitably drilled, and the oil travels through a small tube from the timing cover into the timing shaft. From there it travels into the crankpin through suitably drilled holes and lubricates the big-end assembly. Surplus oil creates an oil mist in the crankcase and this lubricates the main bearings, piston and cylinder and all other moving parts. There are also holes in the crankcase

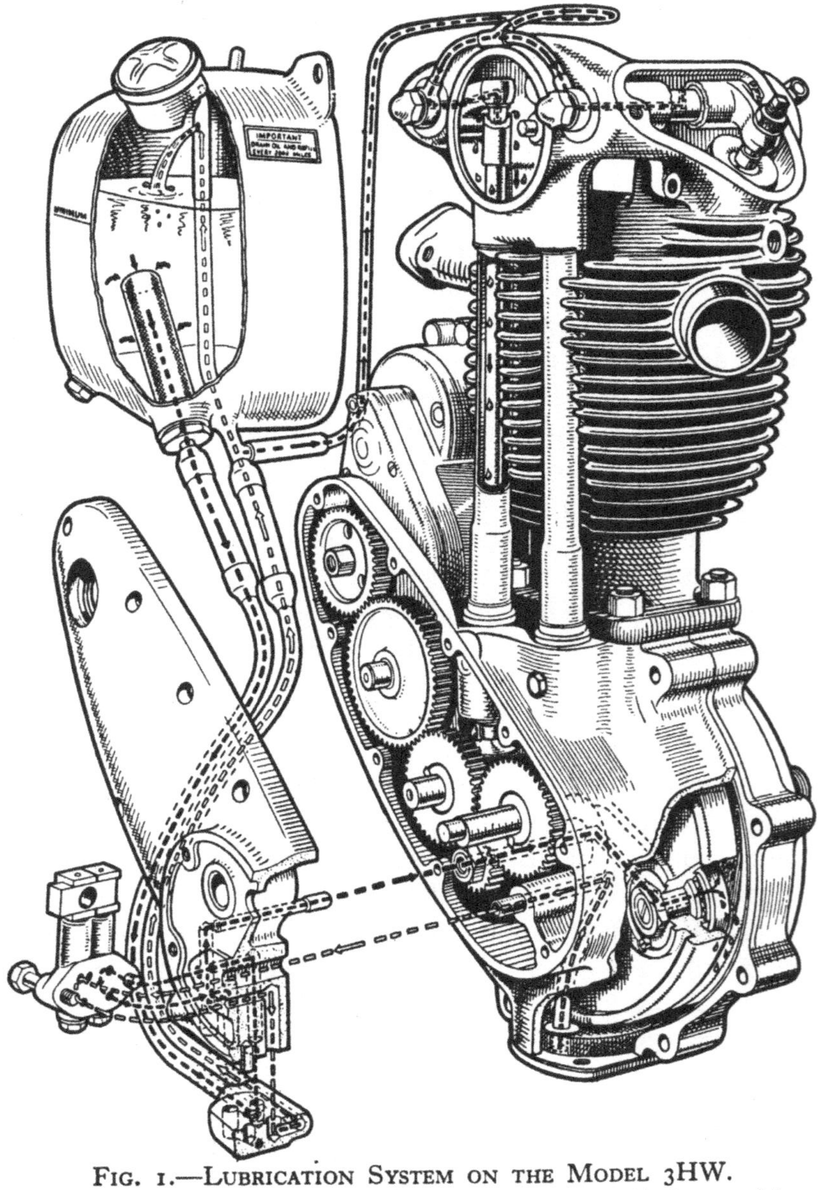

FIG. 1.—LUBRICATION SYSTEM ON THE MODEL 3HW.
The lubrication is the same on all O.H.V. single-cylinder machines with the exception of the feed to the rocker gear.

side of the timing chest which enable the oil mist to pass through to the timing gear. Surplus oil runs down the walls of the crankcase and drains through a filter into the crankcase base. From there it is picked up by the scavenge pump through a return pipe, the intake being at the bottom of the crankcase. The scavenge pump returns the oil to the oil-tank.

Rocker-Gear Lubrication

On the model 3HW a branch pipe from the oil-return pipe is connected to the rocker oil-feed pipe which feeds the two rocker spindles. Oil passes into the drilled spindles and through suitable " bleed holes " into the rocker bushes, from where it leaks into the rocker-boxes to lubricate the moving parts. Surplus oil drains down the push-rod tubes to the crankcase. There is no adjustment to this oil feed, as the quantity fed to the rocker-gear has been found in practice to be adequate, but not excessive, at all times.

On other single-cylinder O.H.V. models the supply of oil for the rocker-gear is taken from the oil-pipe block. A hole connecting with the pressure side of the oil-pump is connected to a suitable pipe which leads the oil up to the rocker spindle by way of an adjuster. There is an adjusting screw and lock-nut so that the quantity of oil can be properly regulated. Another pipe in the same assembly is connected with the oil gauge in order to provide a pressure reading.

Oil-Pump

The oil-pump on all Triumph machines is of the double-plunger type and is actuated by a peg mounted on one of the camshafts. The pump is accessible after removing the oil-pump cover on single-cylinder models, or the timing cover on twin machines. The pump body is secured with two studs or bolts to the crankcase, and, in the case of the single-cylinder models, these are wired together for security. Between the oil-pump body and the crankcase there is a paper washer. The small diameter plunger supplies the pressure feed and the larger diameter the scavenge. Because the scavenge plunger is the larger there is always more scavenge capacity than feed, and consequently the crankcase is kept free of accumulated oil. For the same reason, the oil return to the oil-tank will be

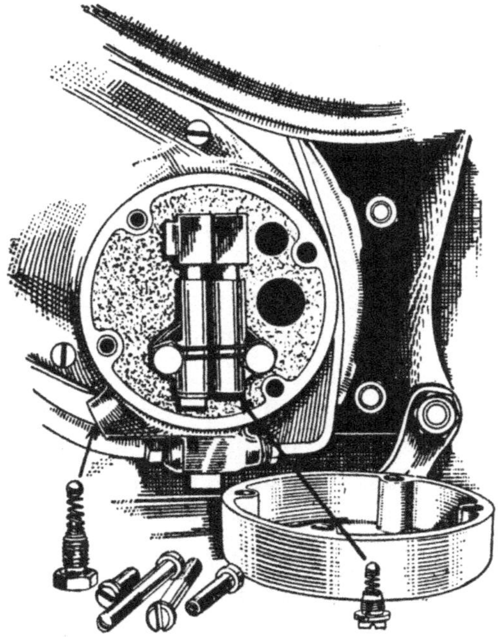

FIG. 2.—THE OIL-PUMP ON SINGLE-CYLINDER MODELS.

Showing the ball-valves and oil-pressure release valve on the left.

spasmodic, consisting of a spurt of oil followed by a number of air bubbles. At the base of each plunger is a non-return valve consisting of a ball and spring. Each valve is closed by means of a screwed plug.

Servicing the Oil-Pump

The plunger type of pump is particularly trouble free and will run for very considerable mileage without attention. A lack of pressure or of scavenge may result from one of the balls failing to seat properly due either to foreign matter or a faulty seating. The remedy is to remove the oil-pump completely, dismantle it and clean all parts thoroughly by washing in petrol. The balls should be very carefully examined, preferably under a magnifying glass, to ensure that they are in perfect condition. After very long service it is sometimes found that the balls are imperfect in shape or have a slight groove in them; if this has happened they should be renewed. The springs should also be renewed if there is any doubt as to their strength. It is unusual to find that the oil-pump body or the plungers have worn, though sometimes the seatings in the body will be found to be defective. The manufacturers can re-cut the seating if the body is sent to them, or a repairer can carry out the operation, but only a minute amount of metal should be removed. The face of the oil-pump body which fits to the crankcase should be carefully examined to see that it is not damaged in any way. When assembling, each ball should be given a sharp tap on to its seating by placing it in position and using a suitable drift. The oil-pump drive block should also be examined to make certain that the centre hole is circular and has not worn oval. If any wear has taken place the part should be replaced. It is essential that a new washer is fitted behind the oil-pump, as a faulty washer can cause serious lubrication trouble.

Lack of Scavenge

Entire loss of or reduced scavenge will mean that oil will remain in the crankcase instead of being returned to the oil-tank. The oil in the crankcase will build up until it starts to be exhausted through the crankcase breather. If there is no scavenge, the oil will soon all be passed out of the breather, and, though this is a very rare occurrence, it is advisable to examine the rear of the machine occasionally, and if it is found very oily to check up immediately on the oil return. Oil will not be returned to the tank if the oil-pump is not working efficiently. At the top of the oil-tank there is a breather pipe. Should this pipe become obstructed, the pressure will build up in the oil-tank and this will prevent an adequate scavenge. Trouble in this direction has also been known to occur owing to the rubber connectors between the oil-tank pipes and the pipes attached to the oil-pipe block being fitted carelessly. If the copper pipes are allowed to dig into the rubber connectors, particles of rubber may be broken off and the lubrication system obstructed. Before fitting these pipes it is advisable to see that their edges are not sharp.

Decarbonising and Valve Grinding

Though it will be found necessary to decarbonise the side-valve engines every 4,000 miles or so, in order to ensure that the unit operates to the maximum efficiency, the O.H.V. models will run for a much greater mileage without this attention. Under normal running conditions at least 10,000 miles can usually be covered with the O.H.V. engine before the head is lifted. The tendency is to disturb the engine too frequently, but it is never advisable to do so unless there are definite signs that this attention is necessary. Poor starting, excessive pinking (on good quality fuel), loss of power with poor hill climbing and acceleration are all signs of a request from the engine for attention. When there is no indication it is not necessary to decarbonise and grind the valves.

General

It is entirely unnecessary to remove the petrol tank in order to decarbonise any Triumph machine. Even with the 500-c.c. O.H.V. engine, there is plenty of room to work with the tank lifted. The removal of the tank entails the disconnection of the wiring from the switch and ammeter and is a waste of labour. In order to lift the tank it should first be drained and the U-pipe at the front disconnected. The instrument panel, where fitted, should now be lifted after removing the screw and the oil-pipe disconnected from the gauge. If the panel screw has not been removed for some very considerable time it may be difficult to unscrew it, but it can be eased with penetrating oil, access to the screw socket being obtained through the hole through which the panel light fits. The petrol-tank-holding bolts should now be removed when, after disconnecting the petrol pipe, the tank can be lifted. It is convenient to take out the saddle-nose bolt and by means of a loop of string on the nose, pull the saddle out of the way, looping the string over the tail lamp. The tank can now be propped up by means of suitable pieces of wood. It is advisable to protect the finish on the top of the tank with a piece of rag to prevent damage by the handlebar arm bolts.

Removing the Cylinder-Head—Side-Valve

The removal of the head on side-valve models is quite straightforward, the cylinder-head bolts and the torque stay only being removed and the head lifted. Before the cylinder-head gasket is removed it should be carefully examined for signs of leakage. A new cylinder-head gasket should always be used before the head is fitted again. The cylinder-head bolts should also be carefully inspected and any which are doubtful as regards the thread or which have damaged hexagons should be replaced. If the head has been "blowing" it is advisable to fit a new set of bolts as a precautionary measure.

Removing the Head—O.H.V.

On all O.H.V. models the head should be removed complete with the rocker-boxes. The procedure is as follows:—

(1) Screw off the carburetter mixing chamber cap.
(2) Remove the petrol pipe.

(3) Take off carburetter, leaving slides attached to the cables. It is convenient to tie these out of the way with a piece of string.

(4) Disconnect the plug lead.

(5) Take off the torque stay.

(6) Remove the exhaust system by slackening off the bolt on the exhaust pipe to cylinder clip, removing the silencer hanger bolt and detaching the pipe and silencer complete.

(7) Disconnect the exhaust-valve lifter cable.

(8) Remove the rocker oil-pipe from the rocker spindles.

(9) Remove the bolts securing the rocker spindle plate to the cylinder. (Models 2H, 3H, T.70 and T.80.)

(10) Remove the cylinder-head bolts and take off the cylinder-head.

Dismantling Cylinder-Head—O.H.V. Models

The rocker-box bolts should be removed and the rocker-box taken off the head (except Army model 3HW). Care should be taken not to lose the hardened end-caps fitted to the valve-stems. It should be noted that the valve-stems have hard stellite tips welded on to them on some later models and all valves on the model 3HW were fitted with these tips. Valve-stem caps are not required where the valve-stems have hardened ends. If there is any doubt, the valve-stem should be cleaned and examined at the end when, if a tip is fitted, it will be possible to see it.

The joint between the rocker cover and the box is made with jointing compound on the 250- and 350-c.c. civilian models and will have to be broken before the rocker spindles are removed. On the 500-c.c. models the spindle can be tapped out from the oil-feed side. On the model 3HW the rocker-boxes are cast integral with the head and the rocker spindles must be tapped out in order to obtain access to the valves.

Note that on all models it is not necessary to dismantle the rocker-box unless some part is to be replaced, but a thorough examination should be made to ascertain whether any wear has taken place. The two operating ends of the rockers in particular should be examined and also the fit of the rockers on the spindles to ascertain if the rocker bushes are not unduly worn. The rocker-box should also be examined for cracks. If the rocker ball pins which fit into the ends of the push-rods show any signs of wear, they should be replaced. The old pins can be tapped out and the new ones driven in. The valves are removed by means of a standard valve-spring compressor. It should be noted that there is a valve-spring seating-washer fitted to the exhaust valve-spring only, the inlet valve-spring seating direct on to the head. The valves are secured by tapered split cotters fitted through a collar.

Removing Cylinder and Valves—Side-Valve Models

Although on side-valve models the valves can be ground without removing the cylinder, this is not recommended, as the job is very difficult to carry out efficiently. To remove the cylinder proceed as follows :—

(1) Remove the exhaust-valve lifter cable.

(2) Take off the exhaust system as described in item (6), page 16.

(3) Remove the nuts from the holding-down studs on the cylinder base. Note that there is a special spanner in the tool-kit to facilitate the removal of the nuts where they are in difficult positions.

(4) Remove the cylinder, taking care that the piston does not fall on to the edge of the crankcase.

The valves can now be removed after taking off the valve-chest cover and using a standard side-valve spring compressor.

Removing the Carbon

There is no better way of removing the carbon than by boiling the parts in a strong caustic soda solution. It must be remembered that soda attacks aluminium and in no circumstances must aluminium parts be immersed. The carbon can be removed by scraping, though this is a more laborious process. Boiling in caustic soda has the disadvantage that the enamel on the head is removed, but this can be ignored, as any efficient worker will re-enamel the head and barrel before fitting as a matter of routine. Care must be taken when removing the carbon from the piston that the crown is not scratched. It is not advisable to remove the piston-rings unless they show signs of gumming, as may occur on side-valve models. When the work of removing the carbon is finished, the interior of the head and ports should be cleaned and polished as far as possible. Where the necessary equipment is available, it is advisable to use rotary wire brushes in order to clean up the ports.

Grinding and Servicing the Valves

The valves and seatings should be carefully examined. On side-valve models cracked seatings should be looked for, and on all models an examination should be made to ensure that the valve heads are not pocketed. Where the seat is too deep to respond to normal cutting, a top cutter should first be used, though this should only be employed in extreme cases. Normally, if the seating is badly pitted, it should be reconditioned by the use of a valve-seat cutting tool, employing the minimum of " cut " to obtain a good seating. It is bad practice to grind a seating if it is deeply pitted. The valve face should be reconditioned in a valve reconditioning tool if in poor condition. Here again extensive grinding should be avoided or the valve will become too deep-seated and performance will suffer.

Valve-stems should be carefully inspected for wear and the valve-guides also. If wear has taken place a ridge will usually be found on the valve-stem and signs of scuffing in some cases. Only very light wear is permissible on the valve-stems, and where wear has developed to any marked degree, the valves should be replaced and the guides also if this should prove to be necessary. In the following section the removal and replacement of valve-guides is detailed. It should be noted that if the valve-guides are to be replaced, attention to the valve-seat should be deferred until this has been done.

After it has been seen that the valve-guides and seats are in good condition and all parts are scrupulously clean, the valves should be ground in. The valves should be placed in position and a little fine valve-grinding compound placed on the face. Grind for a few moments only and then remove the valve, wash in paraffin and examine. If the valve and the seating have been properly serviced, there should be contact all round, and it is only necessary to continue the grinding in order to obtain a perfect joint. On the side-valve models there is a slot in the valve head which enables a screwdriver to be used for grinding purposes. On the O.H.V. models it is necessary to use a valve-grinding tool which grips the stem of the valve, or one of the special suction valve-grinding tools which grip the head of the valve. These are obtainable from the Triumph Spare Part Department. Care should be taken not to grind for too long in one position. The valve should be rotated for a few moments and then lifted off the seating and rotated before proceeding. This makes certain that the small particles of valve-grinding compound do not make a track on the valve face and seating. If the valve and seating have been properly prepared it should never be necessary to use coarse grinding compound. The parts should be periodically washed in petrol or paraffin and the progress of the work examined. When it is considered that the grinding is finished, pencil marks should be made across the face of the valve and this should then be fitted on to the seating and rotated a few times. If the grinding has been satisfactory, the pencil marks should all be removed.

Before the valves are fitted they should be thoroughly washed in petrol or paraffin, and the seating and adjacent parts should also receive the same treatment. If any valve-grinding compound is left behind, excessive wear will occur in the engine. After cleaning, all parts should be lubricated and great care taken to see that the cotters are properly seated and secured. It should be noted that the inlet and exhaust valves differ and that the heads are marked " IN " and " EX ". Care should be taken to ensure that they are properly replaced.

Removal and Replacement of Valve-Guides

The removal of a valve-guide is easily accomplished by means of a shoulder drift. The illustration shows the drift necessary for the various models. It is very important that the shoulder is machined so that it makes a true right angle with the shank. If this angle is not correct there is great danger that the valve-guide will be damaged. With the correct drift, the guide should be tapped out and a new one fitted by the same method. Care should be taken to ensure that the guide is tapped fully home and that all the parts are clean.

FIG. 3.—DRIFT FOR REMOVAL AND FITTING OF VALVE-GUIDES.

The small diameter should be the same as the valve-stem for the model in question.

Assembling the Rocker-Boxes

The rockers are not interchangeable. The exhaust rocker can be recognised, as it has a horn on the end to take the valve lifter. It is not possible to fit a rocker in an incorrect position. On the 250- and 350-c.c. O.H.V. models, except the 3HW, there is a thrust washer inside the rocker-box at each side of the rocker. On the 500-c.c. models there is a rocker return spring which is fitted inside the rocker-box on the drive side and a thrust washer fitted between the rocker-box and the rocker on the timing side. On the model 3HW

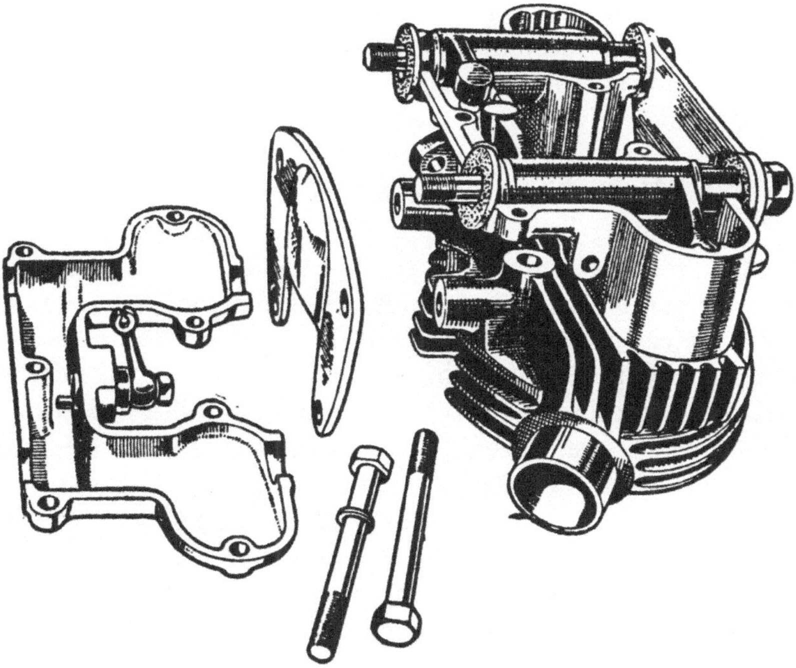

FIG. 4.—THE 250-C.C. AND 350-C.C. SINGLE-CYLINDER HEAD WITH DETACHABLE ROCKER-BOX.

each spindle is fitted with five washers. The rocker spindle should be put through the hole in the side opposite to the oil feed. First place a plain washer, then a spring washer, then another plain washer on the spindle inside the box. Now pass the spindle through the rocker and finally place another plain washer on the spindle before pushing it through the box on the oil-feed side. Finally, tap in the spindle until it is right home. The order of the parts inside the rocker-box from the side opposite the oil feed should be as follows: plain washer, spring washer, plain washer, rocker, plain washer. Where the rocker-boxes have separate covers, the joints should be carefully made with jointing compound after removing any old compound which adheres to the casing. The model 3HW employs gaskets on the rocker-box covers. Where detachable rocker-boxes are employed, do not fit the box to the cylinder-head, as this is more conveniently installed after the cylinder-head has been fitted to the cylinder barrel.

Cylinder-Head Joint

The cylinder-head is ground to the cylinder barrel on O.H.V. models to make a gas-tight joint. No gasket or jointing compound is employed. See that the faces of both parts are clean and apply a very thin, even film of fine valve-grinding compound to the barrel face. Now place the head in position and grind the two faces together until both show an even grey surface. Finally, clean away all traces of grinding compound from both parts. On side-valve models a copper asbestos gasket is used and the old gasket should never be refitted. Great care should be taken to see that the new gasket is not damaged in any way. Before fitting to the cylinder, the gasket should be well covered with engine oil.

Replacing the Cylinder—Side-Valve Models

Before the cylinder is replaced it should be seen that all parts are clean and well lubricated with engine oil. The gaps on the piston-rings should be placed as far away from one another as possible and the cylinder barrel then lowered over the piston. It is convenient to get an assistant to help with this job as it is difficult to hold the heavy barrel and at the same time fit the rings into the bore. A new base-washer should always be used. It should be well covered with engine oil before the cylinder is fitted. After the cylinder is in position, the nuts should be placed on the studs and the engine turned over a few times by hand to see that everything is in order. Finally, the nuts should be tightened up evenly and the assembly is then ready to receive the cylinder-head.

Replacing the Cylinder-Head

The fitting of the cylinder-head on the side-valve models is a job which presents no difficulty, but unless it is properly carried out there may be a leakage which will in time cause burning of the gaskets. After it has been seen that the surface of the head is quite clean, it should be placed in position and the bolts inserted and pulled down gently. The process of tightening down the cylinder-head should now be completed and this should be done by gradually tightening down the bolts, alternating from side to side and end to end of the head. The aim should be to tighten each bolt down just a little at a time in order to distribute the pressure on the gasket evenly. There should be a final tightening down when the engine is

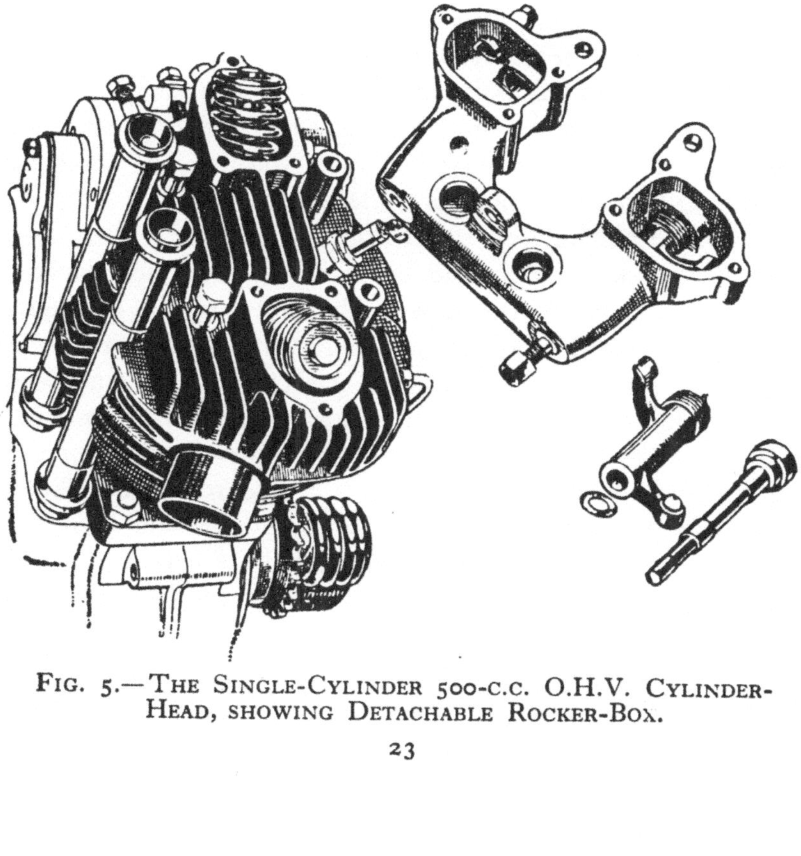

FIG. 5.—THE SINGLE-CYLINDER 500-C.C. O.H.V. CYLINDER-HEAD, SHOWING DETACHABLE ROCKER-BOX.

hot, and the bolts should be checked up again after the engine has been run for a few days.

On O.H.V. models new oil-retaining washers should always be employed at the top and bottom of the push-rod tubes, on the rocker spindles, between the rocker-boxes and the cylinder-head and on all the oil-pipe connections. It is never advisable to use the old washers again, as new ones cost very little and ensure oil-tight joints. Except in the case of the model 3HW, the fitting of the cylinder-head is the same as with the side-valve models, care being taken to tighten down the bolts evenly.

With the model 3HW, the rocker-box and the cylinder-head are all in one piece and the push-rod and push-rod tubes must be fitted at the same time. Before installing the push-rods, they must be carefully examined to see that they are not unduly worn. On this model, the round cover on the head can be removed to enable the push-rods to be properly located. Assembly is assisted if the engine is rotated until the tappets are both in the lowest position. Great care must be taken to see that the push-rod tube washers are correctly positioned in order to prevent oil leakage.

On the other models, after the cylinder-head has been fitted, the rocker-box should be installed, using new rocker-box gaskets between the face of the box and the cylinder-head. In earlier models jointing compound only was employed, but it is better to fit the gasket which was installed on later machines. The gasket should be placed in position, well greased to fix it in the proper position on the head before the rocker-box is put on top of it. The rocker-box bolts secure the head to the box and it should be seen that the push-rod tubes are carefully fitted and the washers properly located. The push-rods should also be carefully examined before fitting to make certain that the ends are in good condition. On the 500-c.c. models, two of the bolts securing the rocker-box also secure the cylinder-head, and the tightening down of the cylinder-head will, therefore, have to be deferred until the rocker-box has been installed. The rocker-box bolts should be tightened down evenly to make certain that the pressure is properly distributed. If this is not done oil leakage may occur.

On the 250- and 350-c.c. models, but not on the model 3HW, there are chromium-plated side plates which secure the rocker spindles. Before these are fitted new rubber washers should be placed over the spindles so as to prevent oil leakage. Each plate is secured by two bolts. On the drive side, nuts are fitted to the rocker spindles in order to secure the rocker plates; on the timing side, the banjo union nuts which secure the oil-pipe serve the same purpose.

Rocker Oil-Pipes

The pipes for the lubrication of the rocker gear are fitted to the rocker spindles on the timing side. The spindles are drilled to take the oil supply and oil under pressure is fed to them. The oil travels through bleed holes to lubricate the rocker bushes; the surplus finds its way to the rocker-boxes where it lubricates the ball pins, push-rod ends and other parts. The banjo union fits over the end of the rocker spindle with a washer at each side to make an oil-tight joint. The whole is secured by means of an acorn nut. These washers are made of copper, and if new ones are not available the old ones should be annealed by heating them to redness and dipping them into cold water.

On the model 3HW the supply of oil is taken from the oil-return pipe just before it reaches the oil-tank. The oil supply is, therefore, not adjustable, but the gauge of the pipe is such that sufficient supply is at all times delivered. With this type of lubrication, it is possible to flood the rockers periodically if the finger is used partly to obstruct the hole in the rocker oil-return pipe in the tank. The engine should, of course, be running at the time. It is a good practice to do this about once monthly to make certain that there is plenty of oil in the rocker-boxes, though the normal supply is adequate to maintain lubrication.

With all other O.H.V. models the rocker supply is taken from the oil-pipe block and is adjustable. An adjuster and lock-nut will be found adjacent to the exhaust rocker spindle in the oil-pipe and with it the quantity of lubricant fed to the rocker-

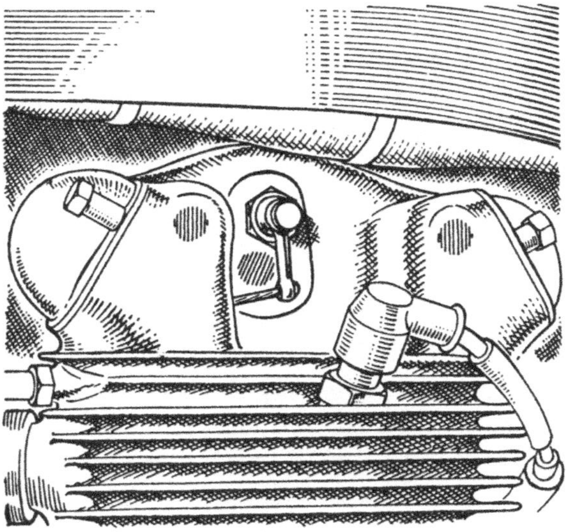

FIG. 6.—VALVE LIFTER FITTED TO THE MODEL 3HW.

The arrangement on other single-cylinder engines is similar.

boxes can be controlled. The rockers are not cooled with oil, but only lubricated, and a very large supply is therefore unnecessary. To test the quantity, the banjo union should be slackened off and levered away from the plate or rocker-box with a screwdriver. The engine should then be started up, and the quantity of oil which leaks away is that which will normally be fed to the rocker-gear. There should be just a slight trickle; if the amount is excessive an adjustment should be made. Too much oil to the rocker-gear will cause oil leakage.

Exhaust-Valve Lifter Mechanism

After attaching the exhaust-valve lifter cable, the adjustment should be checked up. It is possible so to adjust the exhaust lifter that the exhaust valve is held off its seating, with lack of power and burning of the valve face. When making the adjustment, first see that the exhaust valve is closed. Now adjust the cable so that there is about $\frac{1}{8}$ in. free movement on the lever at the engine end. This applies to both O.H.V. and side-valve models, and will make certain that the valve operation is not being interfered with. Quite apart from the danger of having too close an adjustment as previously mentioned, a tapping noise in the engine is sometimes traced to the fact that the exhaust-lifter operating peg is touching the moving part of the valve-operating gear.

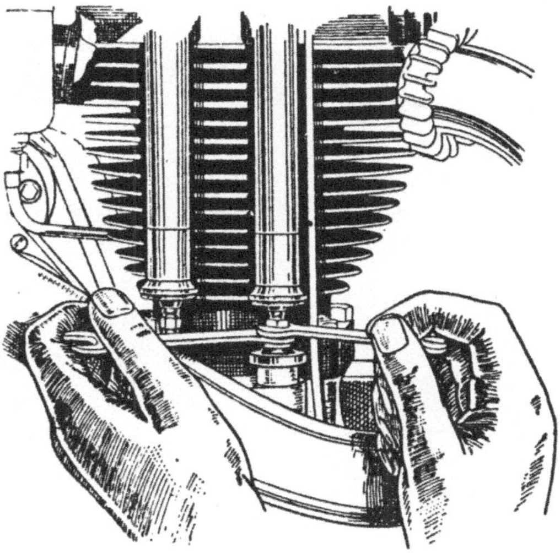

FIG. 7.—ADJUSTING THE TAPPETS ON THE SINGLE-CYLINDER MODELS (EXCEPT 3HW).

Tappet Adjustment

On all Triumph machines, the tappets should be adjusted with the engine cold. On all side-valve models adjustment is made after removing the valve-chest cover. It will be found that there is a lock-nut which locks the tappet-adjusting head to the tappet itself, and this must be slackened off before the tappet head can be moved up or down. On these models, the clearance should be carefully measured by means of a feeler gauge. On O.H.V. models, except on 3HW models, the adjustment is made by lifting the lower portion of the push-rod tubes. The push-rod tubes are in two pieces with a spring in between, and the lower portion fits inside the upper portion. The lower portion can be lifted with a screwdriver and propped up with either a suitable distance piece, or, on all models except the 250-c.c., by resting the flange on the push-rod tube on the cylinder fin. The tappet is then adjusted in the same way as the side-valve models. Care should be taken not to damage the rubber oil-seal washer when making this adjustment. It should be noted that the springs should be strong enough to hold the bottom portion of the push-rod tubes firmly on the seal washer, but if the spring has weakened it should be renewed, which, of course, entails the removal of the rocker-box.

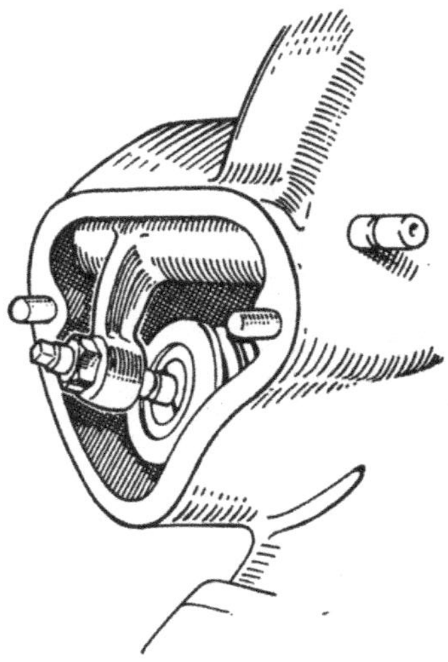

FIG. 8.—THE 3HW ROCKER-BOX WITH COVER REMOVED.

The adjustment is made on the square adjuster and locked with the lock-nut.

On the model 3HW the tappet adjustment is made by removing the covers from the rocker-boxes. An adjustment can then be made on the adjusting pin, which has a square end and is secured by a lock-nut.

On all models it is a good plan to make the adjustment on one valve when the other is fully open. First of all rotate the engine until the exhaust tappet has risen to its full height or, on the model 3HW, the exhaust rocker has depressed the valve to the maximum. Now make the adjustment on the inlet valve, then turn the engine over until the inlet valve is fully open and adjust the exhaust. This is an easier method than putting the piston on to top dead centre before making the adjustment. Many owners of machines spend far too much time adjusting the tappets which normally only require attention at long intervals. The tappet adjustment will not alter unless some parts have worn, and it is, therefore, entirely unnecessary to check up every few hundred miles.

Torque Stays

It should be seen that the torque stay is correctly assembled and, where this is adjustable, put into a state of tension. This part anchors the top of the engine securely and is just as important as the engine-plate bolts, though it is frequently much neglected. Though there are a number of engines which will run satisfactorily without a torque stay, they are usually improved if one is fitted, providing that it is correctly installed.

Tuning and Test

The carburetter should now be fitted after thoroughly cleaning, as described on page 149, and the sparking-plug should be installed after the gap has been set to 0·018 in. Remember that many engines suffer from inferior performance due to the fact that the sparking-plug is in poor condition. If there is any doubt, a new sparking-plug should be installed. Now start the engine and allow it to warm up for about a minute and then adjust the slow running. It will be found that if the adjustment for the slow running is made with the engine thoroughly hot, it may stall when the engine is cold, so a compromise should be aimed at. If good slow running cannot be obtained, various points mentioned in this connection on page 140 should be attended to. With the engine warm, a check-up should be made over all the various nuts and bolts and connections, and the machine should then be taken on the road for a short test. Back in the workshop the cylinder-head bolts should be tightened down again and a check-up made to see that the carburetter joints and those on the petrol pipe are not leaking. Look also for any signs of oil leakage, and check up on the rocker lubrication, as previously described, now that the oil is hot.

Removing Engine from Frame

The removal of the engine from the frame is quite a straightforward procedure and does not present any difficulty. Step by step the procedure is as follows :—

(1) Drain oil-tank.

(2) Disconnect plug lead.

(3) Drain and lift petrol tank as described on page 15.

(4) Disconnect dynamo cable and earth wire on magneto.

(5) Take off advance and retard control from magneto and disconnect exhaust lifter cable.

(6) Remove carburetter (see page 149).

(7) Remove horn, when fitted to front engine plates.

(8) Release oil-pipe block under timing case by taking off nut and pushing block downwards off stud.

(9) Disconnect rocker oil-feed pipe on O.H.V. models, oil-gauge pipe on S.V. models.

(10) Remove torque stay.

(11) Take off exhaust pipe and silencer as described on page 16.

(12) Separate chaincase and remove both halves. This will necessitate the removal also of the clutch (see page 91). The engine shock absorber must also be taken off. The nut has a right-hand thread, and a robust box spanner must be employed, a special tool being obtainable from the manufacturers. It is not necessary to lock the transmission if the tommy bar is given a few sharp blows with a hammer in order to start the nut. The plug with a screwdriver slot in the centre of the nut, where fitted, is part of the engine mainshaft breather and is not a locking screw. This breather was fitted to earlier models, but was discontinued later and a blind undrilled nut substituted.

(13) Remove the front engine plates and undershield, where fitted.

(14) Take out the rear engine plate to frame bolts.

(15) Remove the engine with the rear engine plates attached to the crankcase.

Dismantling

Having removed the engine from the frame, it should be mounted on the bench so that the work on it can conveniently be carried out. Fig. 20, page 60, shows a useful and simple method of holding the engine firmly on the bench. If the vice is a light one, it is advantageous to fit a distance piece in the jaws at the opposite end to that which is holding the crankcase lug. This arrangement has the advantage that there is no heavy jig required to hold the engine, and the fitting shown in the illustration is very easy to improvise.

FIG. 9.—THE CRANKCASE FILTER UNDERNEATH THE ENGINE ON THE SINGLE-CYLINDER MODELS.

Before starting to dismantle the engine it should be thoroughly cleaned externally. This can be done by scraping off the dirt and washing down with paraffin or by means of one of the proprietary de-greasers which are to be obtained from any accessory dealer. Dismantling should now proceed as follows :—

(1) Remove the magdyno (see page 42).
(2) Remove the four securing screws and take off the oil-pump cover.
(3) Cut the security wire on the oil-pump studs and remove them and the oil-pump complete.
(4) Remove the remainder of the screws securing the timing cover and take it off.
(5) Remove the nut from the mainshaft (left-hand thread).

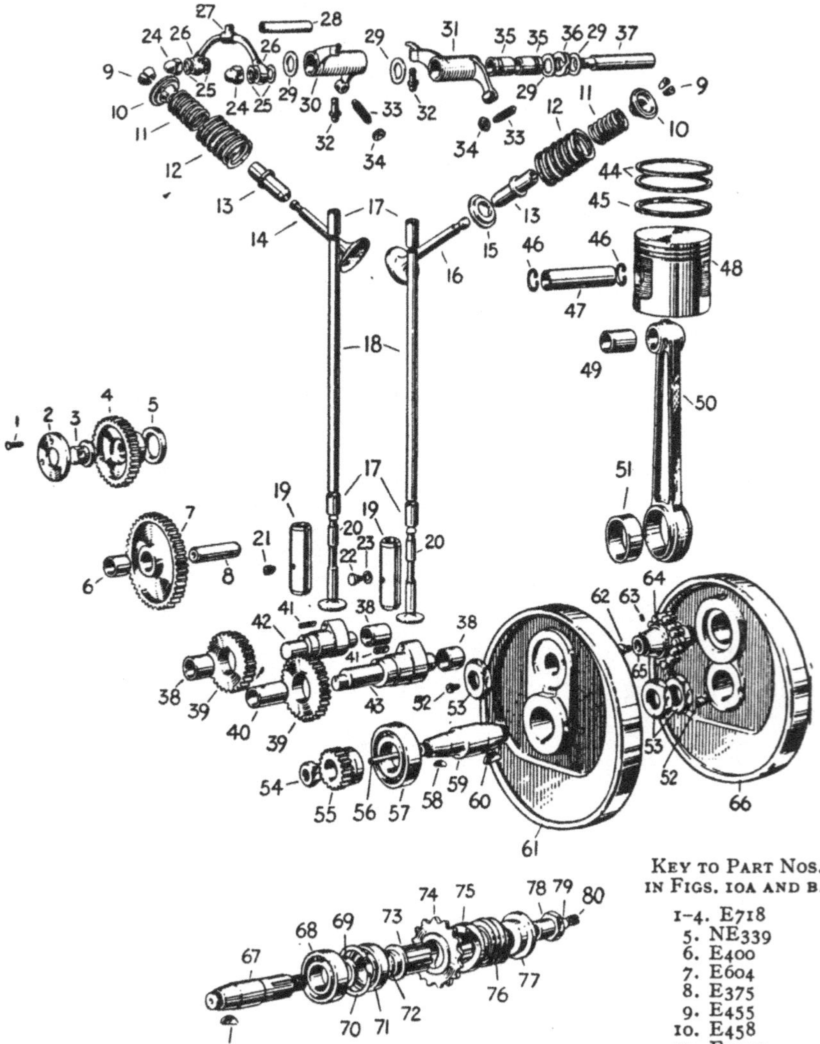

Fig. 10A.—Exploded View of 3HW Engine (See also Opposite Page).

The design of all Triumph single-cylinder engines manufactured since 1936 is similar.

Key to Part Nos. in Figs. 10A and B.

1–4. E718
5. NE339
6. E400
7. E604
8. E375
9. E455
10. E458
11. E1290
12. E1291
13. E303A
14. E2116
15. E1108
16. E2115
17–18. E2119
19. E2105
20. E2104
21. E1413

Key to Part Nos. in Figs. 10A and B (*continued*).

22. T401A
23. E1613
24. E1435
25. E1335
26–27. E2128
28. E1947
29. E2112
30. E2117
31. E2118
32. E949
33. E1513
34. E470
35. E2136
36. E2110
37. E2109
38. E609
39. E499
40. E500
41. E506
42. E602
43. E603
44. E1881
45. E387
46. E361
47. E344
48. E1880
49. E1403
50. E1401
51. E1402
52. E829
53. E398
54. E526
55. E498
56. E1682
57. S35/14
58. S66/10
59. E601
60. KC20
61. E1295
62. E392
63. S31/20
64. TE10
65. E595
66. E1296
67. E600
68. S35/11
69. E510
70. E608
71. S35/14
72. E1732
73. E1521
74. E1680F
75. E1681
76. E659
77. E396
78. E509
79. NE154
80. NE153A
81. E531
82. E530
83. E535
84. E534
85. E538
86. E536
87. E533
88. S26/2
89. S1/51
90. E537
91. E519
92. E532
93. E447
94. E1395
95. E1198
96. E1131
97. F879

(*Continued on next page.*)

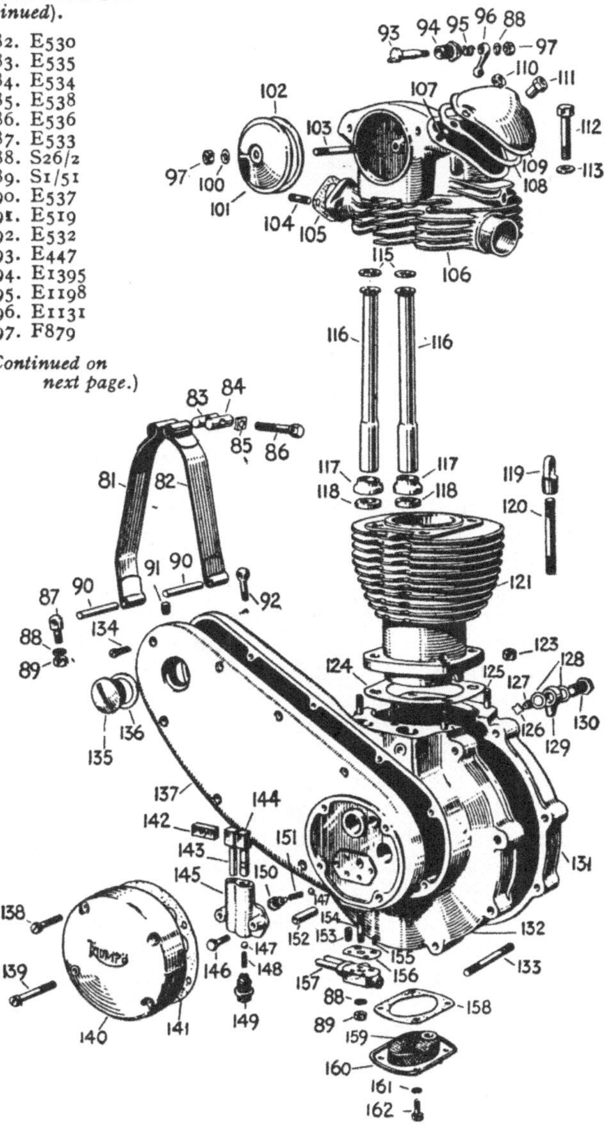

Fig. 10B.—Exploded View of 3HW Engine.

Key to Part Nos. in Figs. 10a and b (*continued*).

100. E2125	111. E2131	126. E1467	138. E427	150. E524
101. E2101	112. E2113	127. E1464	139. E573	151. E1424
102. E2121	113. S26/1	128. E1465	140. E485	152. E1187
103. E2122	115. E2111	129. E1556	141. E490	153. E1100
104. E1396	116–117. E2137	130. E1466	142. E495	154. E416
105. NA43	118. E2114	131. E1350	143. E1354	155. E1143
106. E2100	119. F1033	132. E1349	144. E494	156. E527
107. T454	120. E2124	133. E654	145. E1353	157. E1385
108. E2123	121. E1133	134. E427	146. E575	158. E487
109. E2101 (IN)	123. E671	135. E732	147. S70/7	159–160. E2126
E2103 (EX)	124. E620	136. NE507	148. E403	161. S26/2
110. W103	125. E672	137. E615	149. E386	162. MT63

(6) Extract the mainshaft pinion from the shaft using a standard two- or three-claw withdrawal tool.

(7) Remove the cylinder-head as described on pages 15 and 16.

(8) Take off the cylinder barrel holding-down nuts and lift off the cylinder barrel. Be careful that the piston does not fall on the crankcase, where it may be damaged.

(9) Take out a gudgeon-pin circlip and push out the gudgeon-pin. If it is found that the gudgeon-pin cannot be removed with a light tap, using a suitable drift, the piston should be warmed a little in order to expand it. Take off the piston and mark the inside front with the tang of a file so that it can be assembled again in the correct position.

(10) Remove the crankcase bolts.

(11) Take the engine out of the vice and remove the crankcase filter. There will be a little oil left in the bottom of the crankcase and this should be allowed to drain into a suitable receptacle.

(12) Split the crankcase by "bumping" the end of the mainshaft on a piece of soft wood or lead on the floor. It will be found that the assembly will easily separate.

(13) Remove the crankshaft assembly from the crankcase.

(14) Warm the timing-side crankcase and drop the case from a height of about 6 in. open side downwards, on to a bench. The bearing will then fall out. Another method is to tap the bearing out, employing a disc of such a size that it will only just pass into the crankcase from the outside, bearing against the bearing race. It will thus be prevented from falling through and the bearing can then be tapped out.

(15) Remove the drive-side inner bearing the same way as the timing side and then the drive-side outer bearing after springing out the circlip. Do not lose the distance piece.

Note.—The 3-IW machines fitted with a pressed steel in place of an aluminium chaincase have a cased felt washer installed between the two drive-side bearings. There is also a distance sleeve which keeps the two bearings apart. When dismantling the crankcase (drive side) the inner bearing should first be removed as described in (14) and (15) above. The sleeve which fits the centre of the cased felt washer can be taken out with the fingers or pliers. The felt washer must not be used again in any circumstances and can, therefore, be handled irrespective of any damage which may be done to it. The outer bearings can then be removed in the same way as the inner bearings.

Dismantling the Flywheel Assembly

The dismantling of the crankshaft and flywheel assembly presents no difficulty. The nuts are locked by locking screws

FIG. 11.—THE CRANKPIN AND MAINSHAFT NUTS ARE LOCKED WITH A LOCKING SCREW AFTER THEY HAVE BEEN FULLY TIGHTENED DOWN.

No washer is necessary.

which must be removed. The same box spanner which is used for the engine-shaft nut can be employed for the crankpin and mainshaft nut. All the shafts and the crankpin fit to the flywheels with a taper and can be tapped out after the nuts have been removed.

After dismantling, all parts should be checked for wear, using the data table on page 202. The main bearings should be replaced if they show any signs of rusting, or if, after cleaning thoroughly by washing in paraffin, there is any appreciable amount of play. The bearing cages should also be examined carefully and the bearing discarded if they are not in perfect condition. Should there be the slightest doubt regarding the condition of the main bearings, they should be renewed, as the dismantling of the engine to replace these parts is a considerable task and one which it should only be necessary to carry out infrequently. Note that it is not always necessary to remove the main and timing shafts from the flywheels, as these can be measured up for wear whilst in position.

Assembly of the Flywheel Unit

First fit the main and timing shaft to the flywheels, tighten up the nuts and secure with the locking screws. Care should

FIG. 12.—FITTING THE CONNECTING-ROD.

Showing the string method of positioning the rollers which should be set in grease.

be taken that the nuts are really tightened up, as these pull the shafts into their taper seatings. Now fit the crankpin to the timing-side flywheel and tighten the nut hard up and fit a locking screw. Put a quantity of grease on the bearing surface of the crankpin and fit the double row of rollers in position. The grease will prevent them falling out. To fit the connecting-rod, use a piece of thin string as shown in Fig. 12. This will bring the rollers into position and allow the big-end of the connecting-rod to pass over them. The string should be used on each row of rollers in turn. Though various special jigs have been offered from time to time to carry out this job, it will be found that the string method is entirely satisfactory. Now fit the other flywheel and tighten down the nut. Put a pair of bearings on the timing-side and drive-side shafts and true up the assembly, making a final check with a dial gauge on the gearshaft. Finally, tighten up the crankpin nut and lock with the lcck screw. After tightening up make a final check with the dial gauge.

As a precautionary measure some oil should be pumped

FIG. 13.—V-BLOCKS AND DIAL GAUGE USED FOR LINING UP THE SINGLE-CYLINDER FLYWHEEL ASSEMBLY.

through the assembly from the timing side, using a suitable oil gun and engine oil. If the oilways are quite clear the oil will be seen to leak out from the sides of the connecting-rod between the flywheels. In no circumstances should the assembly be proceeded with unless the oilways are proved by this method to be unobstructed. If the oil will not pass through, the assembly should be taken down again and checked up.

It must also be ascertained that the connecting-rod is quite free after the assembly has been tightened up fully. If the rod is pinched, or the side-play is insufficient, the assembly must be separated again and some metal taken off the flywheel bosses. The correct amount of side-play is from 0·003 to 0·015 in., but with a set of new parts the assembly should, if possible, be given the minimum clearance.

Reassembling the Crankcase—Tappet Guides

Wear on the tappet guide is unusual under normal conditions. In order to remove the guide, the tappet nut should be taken off and the tappet allowed to drop as far as it will go into the crankcase. The guide can now be removed by using a split tube which can be fitted round the shank of the tappet and tapped out with a drift and hammer, using the drift on the foot of the tappet. The split tube must be of a diameter slightly less than the guide and a little longer. It is of assistance if the crankcase is warmed. The tappet-guide fixing-screw must, of course, be removed from the crankcase before the operation is started.

FIG. 14.—THE BACK PORTION OF THE TIMING CASE IS FITTED WITH A CORK WASHER THROUGH WHICH THE MAGDYNO SPINDLE PASSES. This prevents oil leakage.

Timing Gear Examination

The camshaft bushes in the crankcase and the timing cover give very long service and do not normally require replacement. If replaced they must be reamered in line by using a reamer on each pair of bushes at the same time. The intermediate gear is fitted with a bush which requires replacing only after a considerable mileage. This bush must be reamered after pressing in. The intermediate gear spindle is a drive fit into the crankcase, and if it is found to be loose, it should be tinned and replaced.

There is a cork washer fitted behind the magneto drive gear to prevent oil leakage. This should be renewed as a precaution when the crankcase is stripped. The magneto gear should be carefully examined prior to assembly. The rivets which attach the plate to the gear occasionally come loose—usually due to rough handling of the nut—and if this happens the gear should be repaired.

Crankcase

To reassemble the crankcase proceed as follows :—

(1) Fit the bearings to the two parts of the case, not forgetting to spring in the bearing circlip securing the outer bearing on the drive side. There is a distance piece between the two bearings on the drive side and this must not be omitted in any circumstances.

(2) Assemble the crankshaft assembly to the drive side of the crankcase.

(3) Assemble the timing side of the crankcase.

(4) Bolt these two halves of the crankcase together with the crankcase bolt. The joint between the two halves of the case is made with jointing compound. See that the faces of both cases are clean before the compound is applied.

Note.—If a pressed steel primary chaincase is fitted on the model 3HW, special attention must be paid to the installation of the felt washer between the drive-side bearings. This washer must be soaked in engine oil for two hours at least before fitting. After fitting the lubricated outer drive-side bearing to the crankcase shell and springing in the retaining circlip, take the cased felt washer out of the oil and press it into the crankcase shell. Place the distance sleeve through the felt washer and install the inner bearing.

Valve Timing

The valve timing is marked and if the marked teeth are meshed together the timing must be correct. When carrying out this operation, the " EX " marking on the inlet camwheel and the " IN " marking on the exhaust camwheel should be ignored. If the markings on the small timing pinion are not visible, the pinion nut should be moved slightly so as to expose

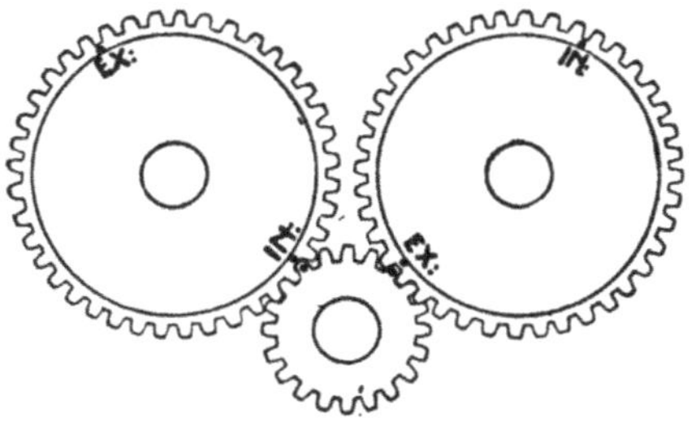

FIG. 15.—THE CORRECT POSITION FOR THE TIMING GEARS ON THE SINGLE-CYLINDER MODELS.

Note the double markings on the cam wheels.

the mark. If the valve timing is to be checked with a degree chart, the valve tappet clearance must first be set to a " Nil " clearance.

Timing Gear Assembly

The assembly of the timing gear is quite straightforward, the small gear fitting on to the timing-shaft taper, the position being secured with a key; the nut has a left-hand thread. The mesh of the various timing pinions can only be checked if a skeleton cover is made for the timing chest. This is a cover with parts cut away so that the gears can be inspected. The teeth should mesh together with only the very slightest degree of clearance. The amount of movement on one tooth between two teeth of its mating gear should be hardly perceptible. If there is any considerable degree of movement, then the timing gear will usually be noisy. On the other hand, if the mesh is too close, the gears will whine. The meshing of the timing gear is an operation which requires to be carried out very carefully if the engine is to run with reasonable silence. A correct mesh cannot be arrived at if the bush or spindle for the intermediate gear is worn or if the camshaft bushes have excessive clearance. It must be remembered, however, that if the camshaft is entered into the bush in the crankcase and the opposite end gripped with the finger and thumb, it will be possible to move it up and down a little. The only accurate method of gauging the wear on the camshaft bushes is by means of a plug gauge.

Piston and Cylinder Barrel

The cylinder barrel can be reconditioned by the fitting of a liner or by re-boring. On the single-cylinder engines the re-bore should not be carried out to more than 0·040 in. oversize. The makers' standard oversizes are 0·020 in. and 0·040 in., and in these sizes pistons and rings are available.

The gudgeon-pin should be a tight fit in the piston bosses. If the pin can be pushed freely into the piston at atmospheric temperature then it is too slack a fit. The piston should be heated a little before fitting the gudgeon-pin, which will make it possible to slide it easily into position. Before installing the gudgeon-pin, it should be seen that it is a free fit in the small-end bush but without any play when the parts are dry and unlubricated. The gudgeon-pin is held in position by circlips which must be fitted carefully to make sure that they are fully engaged in their grooves. If a circlip is omitted or insecurely fitted the gudgeon-pin is likely to groove the cylinder.

If new piston-rings are fitted they should be tried in the barrel at the bottom of the bore and the gap measured. The piston should be used to push the ring up the barrel, as this will make sure it is correctly positioned in the bore. The gap should be measured with a feeler gauge. Each piston-ring should also be tried in its groove to make certain that it is free and also that the groove is not worn, thus giving an excessive clearance. If the same piston-rings are to be used as were fitted previously, it should be seen that they are quite free in the grooves and not gummed up or obstructed with carbon. If this has happened, then the rings should be removed and thoroughly cleaned.

After the piston has been fitted and well lubricated, the cylinder should have attention. If the cylinder has not been re-bored and there is a step at the top of the bore caused by wear, this should be removed with a scraper, otherwise the top piston-ring may be damaged. If there is any sign that the piston has at any time seized in the bore, care should be taken to remove any aluminium " smear " which may remain in the cylinder wall. This can be carefully removed with a scraper and finally polished with a piece of well-worn emery cloth.

When dealing with an engine the history of which is not known to the fitter, care should be taken to see that the correct size of piston is fitted should a new one be installed. Normally, it will be found that the oversize, if any, is stamped on the crown of the piston. The possibility that the cylinder has at some time been re-bored should also not be overlooked.

Fitting the Cylinder and Cylinder-Head.—This operation has been fully described in the decarbonising section and reference should be made to pages 22 and 23.

Removing the Magdyno

In order to remove the magdyno it is not essential to take off the timing-case cover. The timing plug at the top of the case should be removed, when the nut securing the gear to the armature shaft is accessible. Unscrew the nut with a slow but firm turning movement (right-hand thread), which will release it from the shaft and will also withdraw the gear from the shaft. The shaft has a key-way cut in it, but no key is fitted.

Remove the two dynamo leads by taking out the centre screw

on the kidney-shaped fixing plate. (Note that it is impossible to refit the leads in an incorrect position). Release the earth wires by screwing out the securing plug adjacent to the contact-breaker cover. Screw out the advance and retard control after turning over the ear of the tab washer. After releasing the clamp on the top, the magdyno can be removed from the engine.

Fitting the Magdyno

The magdyno should be installed on its platform and clamped into position. Before thoroughly tightening up the clamp the magdyno should be levered away from the timing case, as this will make certain that the maximum amount of pressure is put on to the oil seal. If originally removed, the timing-case cover should now be installed, making the joint with jointing compound. Make certain that both faces are quite clean before the compound is applied. It is better not to fit the oil-pipes and block until the engine is back in the frame, as this makes the job of connecting up with the rubber connectors to the oil-tank pipes easier.

The timing of the magneto is fully described in Chapter VIII on pages 200 and 201. Having completed the timing, the sparking-plug can be fitted and the lead attached.

Replacement of Engine in Frame

The replacement of the engine in the frame is the reverse of the removal operation. It will be found convenient to leave the bolts which attach the rear engine plates to the crankcase loose, as this facilitates the location of the engine plate to frame bolts. Make certain that every nut is tight, particularly the engine shock-absorber nut and the clutch nut on the gearbox mainshaft. A new washer should be employed for the primary chaincase joint. The joint between the chaincase and the crankcase is made with jointing compound.

Starting Up and Testing

The engine can now be started up and a check should at once be made to see that oil is returning to the oil-tank and the oil gauge, where fitted, is registering not less than 5 lb. pressure. A final test should be made as described on page 30.

CHAPTER II

TWIN ENGINES

THE "Twin" engines are the two-cylinder vertical type produced in three capacities: 350-c.c., 3T–3T De Luxe; 500-c.c., 5T–Speed Twin, T100–Tiger 100, TR5–Trophy; and 650-c.c., 6T–Thunderbird.

The cylinder is of monobloc construction and is fitted with a detachable cylinder-head. The cylinders, blocks and heads are of special-quality cast-iron for all models, except the Trophy and 1951 Tiger 100 which employ aluminium alloy castings fitted with cast-iron liners. The cylinder-head has four valves fitted, two to each combustion chamber, valve-tips being stellited to minimise wear. All valves are independently operated by rockers through short push-rods driven from the camshafts which are situated fore and aft in the crankcase. Rocker-boxes are detachable on all models except the 350-c.c. in which the rocker-boxes are cast with the cylinder-head.

The 500–650-c.c. balanced two-throw crankshaft is a built-up assembly, comprising two crank webs with the journals and a central flywheel, the whole assembly being held together with six 55-ton steel bolts and nuts. Connecting-rods are "H"-section forgings of RR.56 alloy and the bottom caps are steel stampings lined with white metal, the two parts being bolted together by two high-tensile bolts. Phosphor-bronze bushes are pressed into the connecting-rod small-ends.

With the 350-c.c. engine, the crankshafts are pressed into the flywheel, the bosses being slotted and drilled to accommodate the pinch bolts which lock the assembly. Plain bearing "H"-section connecting-rods forged from steel are fitted to the crankshaft before assembly as the big-ends are not split.

The big-end is lined with white metal and the small-end has a phosphor-bronze bush pressed in.

Cast aluminium low expansion light alloy pistons are employed and have two compression rings and one scraper ring. The hardened steel gudgeon-pin is a floating fit in the small-end and an interference fit in the piston, the pin being positioned by two circlips.

The crankcase is cast from aluminium alloy and is designed to give the utmost rigidity. This part houses the crankshaft, timing gears, oil-pump and camshafts. Spur-type timing gears are used and from these the valve gear, oil-pump and auxiliaries are driven, the magneto being fitted to the rear of the crankcase and the dynamo at the front. Early models are fitted with a magdyno only at the rear.

Carburation is by the Amal carburetter which provides mixture for both cylinders.

A B.T.H. or Lucas magneto is used for ignition and is driven at half engine speed off the inlet camshaft gear. The magneto is the two-cam lobe-type, with the cams set at 180 degrees apart. Manual control was fitted to all 1938/40 models and was later re-introduced for the Trophy and 1951 Tiger 100. All other models manufactured since 1945 are fitted with an automatic device which sets the spark in relation to engine speed.

Lubrication of the engine is of the dry-sump type, delivery scavenge being carried out by a double-plunger pump which operates off the eccentric spindle on the inlet camshaft gear.

Lubrication of Engine

All Triumph engines are lubricated by the dry-sump system. Oil is fed to the oil-pump by suction and gravity from the oil-tank, through a filter attached to the large hexagonal nut located at the bottom of the tank. When the oil issues from the pressure side of the oil-pump, it first passes

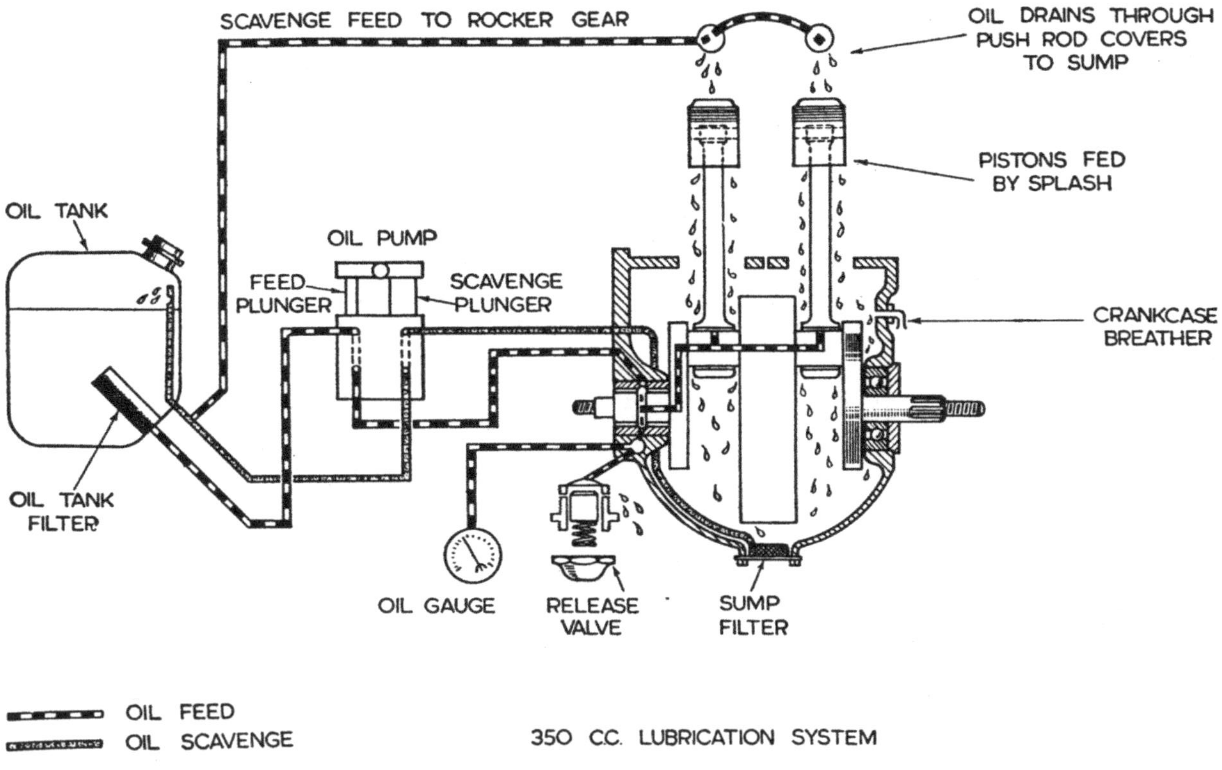

FIG. 16.—LUBRICATION SYSTEM ON 350-C.C. ENGINES.
In 1949 the oil-gauge was eliminated and an indicator fitted to the pressure-release valve.

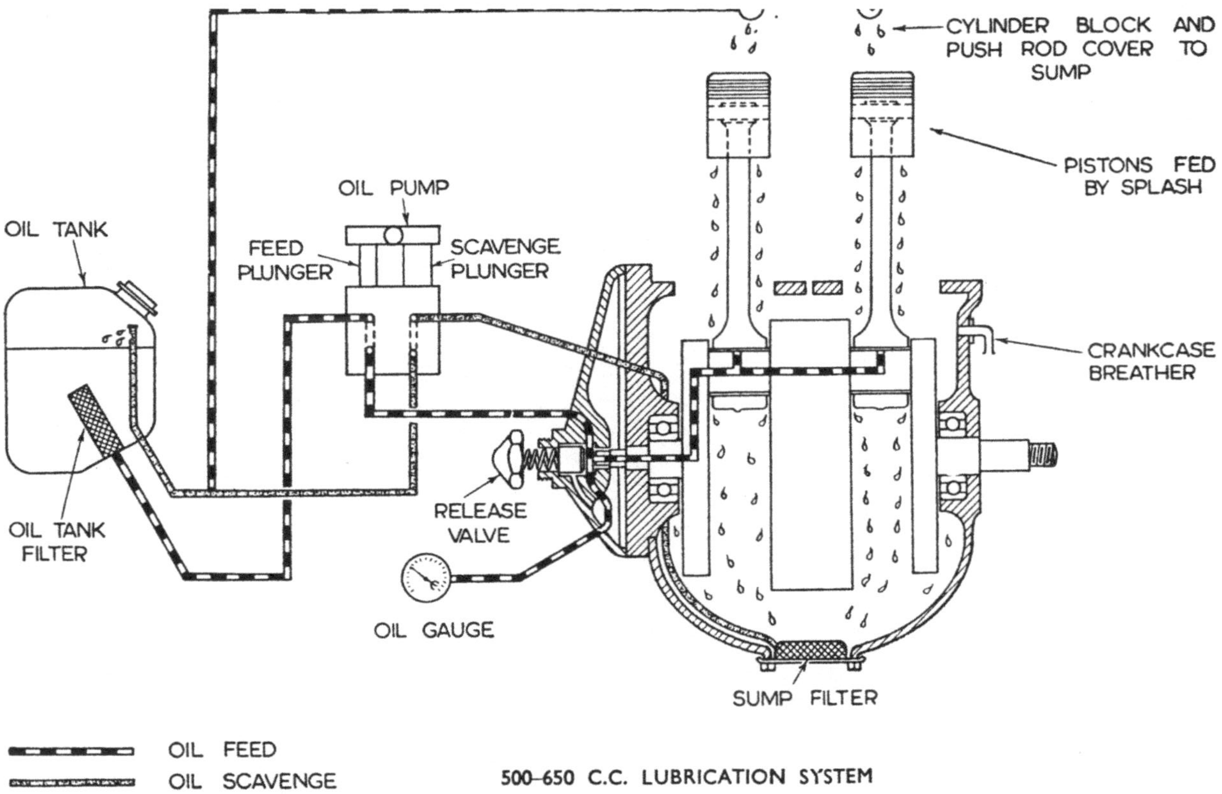

FIG. 17.—LUBRICATION SYSTEM ON 500-C.C. ENGINES.

through drilled passage ways in the crankcase to the front of the crankshaft bush in the timing cover (500-650 c.c. model). From here, oil is directed to the crankshaft and pressure indicator or gauge on the 1938/40 model to the rocker gear also, excessive oil pressure being by-passed by the oil-pressure relief valve to the sump and then scavenged back to the oil-tank.

The passage of oil differs slightly with the 350-c.c. model. The pressure oil from the pump, after passing through the drilled passage ways, is directed to an annular groove in the timing-side main bearing. The oil is then picked up in a corresponding groove in the timing-side gearshaft and is fed direct to the crankshaft, excess pressure of oil being released by the oil-release valve. After the oil passes through the crankshaft to the big-end bearings, it leaves the bearings in the form of an oil fog. This fog lubricates the main bearings, pistons, cylinder block, cams, tappets, and also passes into the timing case to supply lubricant for the timing gear. The surplus oil runs down the walls of the crankcase and drains through an oil filter into the crankcase base. From this point the oil is scavenged back to the oil-tank by the oil-pump. The scavenge oil return pipe will be seen protruding through the oil filter after the cover plate has been removed.

Lubrication of Rocker Gear

The oil feed to the rocker gear is taken from the scavenge side of the oil system. The return pipe in the oil-tank is tapped and a branch pipe is connected to feed to the rocker spindles. Oil passes into the drilled passage ways of the spindles and so lubricates the valve rockers. The rockers are also suitably drilled to carry oil to the ball pins, thereby giving full lubrication to the push-rod cups. On the 500–650-c.c. (1938/40 and 50/51) machines surplus oil drains down external pipes to the push-rod cover tubes, whereas the 1945/49 models have internal drilled drainage-oilways in the cylinder-heads and blocks. Oil issuing from rocker ball pins drains into the push-rod cover tubes lubricating the lower push-rod cups and tappet stems, and finally draining into the crankcase via the tappet guides. With the 350-c.c. model the oil is fed to the rockers and push-rods in exactly the same manner, but drains down through the push-rod cover tubes only.

On 1938/40 500-c.c. Twin Models, the oil is fed to the rocker gear from the pressure side, early models having a common pipe from the timing case to the rockers and oil gauge. Method of adjustment is by restrictor taper screw. The modified type has twin pipes from the timing case, one direct to the oil gauge and the other to the rocker gear. Flow of oil to the rocker gear is determined by a restrictor fitted in the body of the oil-pipe connection at the timing case. With this type of restrictor no adjustment is necessary.

Oil-Pump

The oil-pump is of the double-plunger type, and is actuated by an eccentric spindle on the inlet camshaft through an aluminium drive block. To gain access to the oil-pump it is necessary to remove the timing-case cover. The pump body is secured to the crankcase timing side by two studs and nuts; to remove the pump assembly it is only necessary to take off the two nuts and slide the pump off the studs. Between the pump body face and the crankcase a thin paper washer is fitted to make an oil-tight joint.

It will be noted that the plungers are of different diameter, the smaller being the pressure feed and the larger the scavenge. As the scavenge plunger is the larger, the scavenge capacity will always be greater than the feed, consequently the crankcase is kept free of accumulated oil. For the same reason the oil return to the oil-tank will be spasmodic, consisting of a spurt of oil followed by a number of air bubbles. In the lower part of the pump two non-return valves are fitted; to remove these for inspection it is only necessary to unscrew the two plugs at the base of the pump.

Oil-Pressure Release-Valve

The oil-pressure release-valve is fitted in the front offside crankcase on 350-c.c. models, and in the timing cover on all other models. The following types have been employed: (*a*) ball valve (1938); (*b*) ball valve modified and filter gauze added (1939/40); (*c*) piston valve (1945/48); (*d*) piston valve modified to incorporate pressure indicator. When a replacement release-valve is required, the (*d*) type supersedes (*b*) and (*c*)

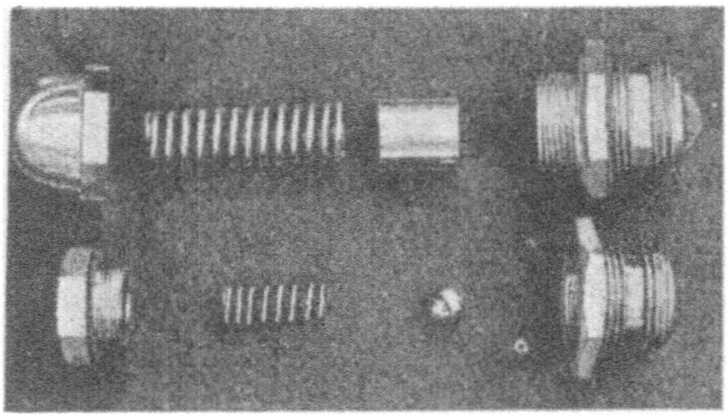

FIG. 18.—THE TWO OIL-PRESSURE RELEASE-VALVES.

The ball type were used on pre-war Twin Models and the piston type on post-war machines.

only. The release spring in each case is set at 80 lbs. per sq. in.; excess pressure will force the valve off its seating and allow pressure oil to be by-passed through the drilled ducts in the release-valve body to the crankcase sump.

To remove assembly for inspection, unscrew unit from its housing. To dismantle, remove outer cap nut and withdraw valve and spring. In the case of type (*d*) further dismantling is necessary. Grip cap nut lightly in a vice and insert screw-driver into recess of indicator button and screw off spring stop nut, the latter being held by a suitable tool. After thoroughly washing, all parts should be carefully assembled. Types (*a*) and (*b*) should have seatings trued by tapping the hardened ball on to them and with (*c*) and (*d*) the piston should move freely in the body of the valve.

Servicing the Lubrication System

The lubrication system employed in Triumph engines is very simple and gives long service without attention. The oil should be changed every 1,500 miles, but in adverse conditions the oil changes should be made at more frequent intervals.

Oil Pressure

Pressure indication on machines dating from 1938 to 1948 is by oil-gauge, but, due to the reliability of the lubrication system, it has been possible, on later models, to dispense with this in favour of a mechanical oil-indicator. The oil-gauge reading should be 35 lbs. per sq. in. at 35 m.p.h. in top gear with the oil hot. With the mechanical type, the rider should observe that the indicator is projecting before riding away; if not, the engine should be stopped and the trouble investigated immediately. The fact that oil is returning to the tank is not proof that it is passing through the crankshaft and lubricating the engine. Only if the pressure is being indicated can the rider be sure that the engine is obtaining sufficient lubricant.

Low Pressure—Causes and Cures

The causes of low oil pressure can be traced to any of the following:—

- Faulty oil-pump.
- Faulty oil-pressure release-valve.
- Old or diluted oil.
- Incorrect grade of oil.
- Worn oil-pump block.
- Worn big-ends.
- Worn bush in timing cover.
- Worn timing-side plain bearing (3T).
- Faulty oil gauge or pipe.

When diagnosing the cause of low oil pressure, the behaviour of the oil gauge is a good indication of where the trouble lies though it cannot be located in all cases by this means. The following information will assist in finding the fault.

(1) *Pressure low* at all throttle openings. Usually due to very thin oil.

(2) *Pressure rises and falls* considerably as the throttle is opened and closed, being greatest at high revs. Suspect faulty oil-pump or release-valve, worn drive block or bush in timing cover.

(3) *Pressure low,* but if machine is ridden at about 40 m.p.h. and throttle then closed, pressure rises considerably. Worn big-end bearings.

(4) *No oil pressure* shown on gauge. Faulty gauge or obstructed pipe. Very badly worn oil-pump drive block.

The following are remedies for low pressure.

Thin Oil.—The oil may be thin due to age or dilution with petrol which is often caused by the use of the machine for very short runs when the engine condition is poor. The use of a very thin oil, particularly of poor quality, will cause a low gauge reading. Only the oils recommended on page 211 should be used.

Faulty Oil-Pressure Release-Valve.—On 1938/40 models.—Remove the valve from the timing cover, being careful not to lose the washer. Dismantle and wash all parts in petrol. Before reassembling place the body on the bench or other firm surface, gauze downwards, and drop the ball on to its seating. With a hammer and a suitable punch give the ball a sharp tap on to its seating. Reassemble the parts and fit to the timing cover. If this treatment shows an improvement in pressure, but the gauge reading is still rather low, the spring in the valve may be pulled out a little in order to increase its strength.

On post-war models the piston valve should be removed and polished and the spring free length checked in accordance with the data sheet.

Faulty Oil-Pump or Drive Block.—Take off timing cover by removing screws and disconnecting oil-pipes. Screw nut back on to oil-gauge pipe connecting-stud and tap gently to assist in removing cover. The joint is made with jointing compound and must be broken. Do not use a screwdriver between the cover and crankcase faces. Remove oil-pump and examine block. The drive hole in the centre should be circular. If worn, the hole will be oval, which indicates that a new block is required. Dismantle the oil-pump by taking out the plungers and screwing out the twin plugs with square heads. A ball and spring will be found under each plug and these, together with the other parts, should be carefully washed in petrol and dried. Now stand the oil-pump body upside down on the bench or other firm surface and place a ball on its seating. By means of a hammer and a suitable punch give the ball a sharp tap on to its seating. Fit the spring and plug and repeat with the other ball. Reassemble the pump and fit to the engine. For further information, see page 12.

Worn Bush in Timing Cover.—Remove the timing cover and, after removing the exhaust pipe, re-fit the bush into the shaft, holding the cover upside down. Now rock the cover to see if the bush is a good fit on the shaft. If not, fit a replacement, part number E.378. The bush should be pressed into the cover after removing the old one and no reaming is required. Note that it is useless to fit a new bush if the engine main bearings are worn, as it will fail again in a very short mileage.

Worn Big-End Bearing.—Slight wear in the big-end of the light-alloy connecting-rods can be taken up by re-bedding (see page 82). As the connecting-rod fitted to 3T models cannot be split, wear in the big-end can be rectified only by re-metalling.

Worn Timing Side Main Bearing (3T).—Wear here allows pressure oil to leak between the bearing and timing-side crank-shaft which either reduces or eliminates the supply to the big-ends (see page 78).

Faulty Oil Gauge or Pipe.—Trouble with these components is very rare. Oil gauges cannot normally be repaired. Replacement oil-pipes and gauges are serviced. An oil-pipe may be faulty by being dented due to accident or to stoppage with foreign matter, though this is unusual. A faulty gauge usually fills up with oil or does not return to zero when the engine is stopped. In either case a replacement is indicated.

Oil Not Returning to Tank

Entire loss of, or reduced, scavenge will mean that oil will remain in the crankcase instead of being returned to the tank. Eventually the oil in the crankcase will be exhausted through the breather valve until the supply of oil in the oil-tank is nil. The first indications of this trouble will be the amount of oil blown back on the rear wheel. If the oil-pump is not working efficiently, oil will not return to the tank (see " Oil-Pump ", page 11). At the top of the oil-tank will be found a vent pipe. Should this pipe become obstructed, pressure will be built up in the tank and prevent adequate scavenge.

Remove the oil feed and scavenge pipe assembly from the tank and apply pressure in the scavenge pipe either by blowing down it or using a pump. This will, of course, indicate any obstruction. It has been found that if the rubber connections have been incorrectly fitted, due to the pipe being allowed to dig into the rubber, a portion of rubber is likely to become detached, thus obstructing the oilways.

If after carrying out these tests the trouble still persists, it will be necessary to dismantle the engine in order to inspect the oil-return pipe in the crankcase. Any air leak at this point will cause complete failure of the scavenge system.

High Oil Pressure

Excessive pressure can be caused by the release-valve spring tension being altered on the piston in the release-valve body or the piston being held in the closed position due to foreign matter. If the fault is the former a new spring should be purchased, and in the latter case the piston should be removed from the valve body and the parts thoroughly cleaned and polished until the movement of the piston is quite free.

DECARBONISING AND VALVE GRINDING

Dismantling—5T, 6T and T.100

(1) On models manufactured during 1949, and after, the petrol-tank can be removed, but with earlier machines it is more convenient to raise the tank, to avoid disconnecting the wiring harness at the instrument panel. To do this, carry out the operation in the following manner: Remove screw securing instrument panel and lift it until union nut on oil-gauge pipe is accessible. The oil-pipe can now be detached in order to obviate any damage when the tank is lifted. Remove four tank bolts and detach petrol-pipe. Take out saddle-nose bolt; tie back saddle by placing a piece of string over the nose and looping it around back of number plate. The tank can now be raised to the required height. Before raising, place a duster over front end of tank to prevent handle-bar brackets scratching enamel.

(2) Disconnect H.T. cables and remove sparking-plugs.

(3) Disconnect torque stays, carburetter and exhaust pipes.

(4) Take off the rocker-gear feed-pipe by removing the two banjo unions at the rocker-spindles and rocker drain pipes where fitted.

(5) Remove the rocker-boxes by unscrewing the four central cylinder-head holding-down bolts, four rocker-box stud nuts and the four screws.

(6) Should it be necessary to remove the rockers from the rocker-box the spindles should be tapped out from the threaded end, using a soft tool. Care should be taken to observe the positions of the rocker-spindle thrust-washers before commencing the operation. Rocker washer order: plain washer,

spring washer, rocker, plain washer, plain washer, rocker, spring washer, plain washer.

(7) Unscrew rocker drain bolts where fitted and remove the front push-rod tube and push-rods, then unscrew the remaining four cylinder-head bolts and lift the cylinder-head off the block. Take away the inlet push-rod tube and push-rods.

(8) To remove the valves from the cylinder-head a special tool should be used to compress the valve-springs. In the absence of this tool, a suitable piece of wood can be made to fit the contour of the cylinder-head combustion chamber. This will prevent the valve itself dropping when pressure is exerted on the spring collar. A piece of hard wood should then be suitably slotted, when it can be pressed down on the valve-spring collar and will expose the cotters, which can be tapped out with a screwdriver. Mark the valves before removal to enable correct re-positioning.

(9) Should it be necessary to remove the cylinder block, care must be taken not to drop the tappets into the crankcase. To overcome this, place a rubber band around the heads of the tappets before lifting. If piston-ring breakage is suspected, raise the pistons to their highest position, lift the block a little and place some clean cloth or two suitable pieces of cardboard over the crankcase aperture. This will obviate broken rings falling into the crankcase when the block is lifted clear of the pistons.

(10) The pistons can now be taken off by removing two circlips and tapping the gudgeon-pins with a soft drift clear of the connecting-rods, care being taken to support the opposite side of the piston while carrying out the operation. Note that pistons must be marked before removal to aid correct reassembly.

Dismantling TR5

(1) Remove the petrol tank.

(2) Disconnect H.T. cable and remove sparking plugs.

(3) Remove torque stays and exhaust pipes.

(4) Take off the carburettor and then the manifold.

(5) Remove the two acorn nuts securing the rocker oil-feed pipe and ease the banjo unions off the spindles.

(6) Unscrew the four rocker oil-drain bolts at the cylinder-head ; leave pipe attached to the cover tubes.

(7) To remove the cylinder-head, unscrew the eight holding-down bolts and remove from the head. Tilt the head towards the rear and remove the front push-rods and cover tube ; the head can now be lifted clear and the rear push-rods removed.

(8) Detach the rocker boxes from the cylinder-head.

(9) For further information turn to page 54, paragraphs 6, 8, 9 and 10.

Dismantling 3T

(1) Disconnect exhaust pipes, carburetter, engine-steady bracket, H.T. cables and remove plugs.

(2) Take off the rocker-gear feed-pipe by removing the two banjo unions at the rocker spindles.

(3) To remove cylinder-head, unscrew the eight holding-down bolts. Lift the cylinder-head vertically and detach the push-rods and push-rod tubes, when the head can be lifted clear of the cylinder block.

(4) To remove the valves from the cylinder-head, a valve-spring compressor tool is required. When the valve-springs are in a compressed state the valve-spring cotters can easily be detached with the use of a screwdriver. Carefully mark the valves to aid replacement.

(5) If the rocker-arms are quite free on their spindle it is undesirable to strip the rocker gear. Should it be necessary, the spindles can be tapped out from the threaded end, using a soft tool, and the rocker-arms extracted from the rocker-box.

(6) To remove the cylinder block it is only necessary to raise it clear of the crankcase studs. Before doing so care should be taken to prevent the tappets falling into the crankcase. This can be done by placing a short ¼-in. bolt between tappet grooves.

(7) Pistons can be taken off by removing two circlips and tapping out gudgeon-pins clear of connecting-rods with a soft drift or suitable withdrawal tool. Mark pistons to aid correct reassembling.

Preparation for Reassembly of Engine

Lay all parts out in their correct reassembly order. Moving parts and surfaces upon which they work should be liberally oiled with the use of an oil-can or brush. Grease the paper and copper gaskets, as this greatly assists during erection of the engine and helps to make a more oil- and gas-tight joint.

Reassembly of 5T, 6T and T.100 Engines

(1) Replace pistons—ensure correct positioning of circlips and that the gaps in the piston-rings are not in line.

(2) Place piston-ring clips over pistons to secure rings.

(3) Prepare cylinder block—grease base washer and fit to block. Fit the four tappets and clip together with a rubber band. The cylinder block can now be placed over the pistons.

(4) Remove piston-ring clips and assemble cylinder block to crankcase; grease and fit cylinder-head gasket to block.

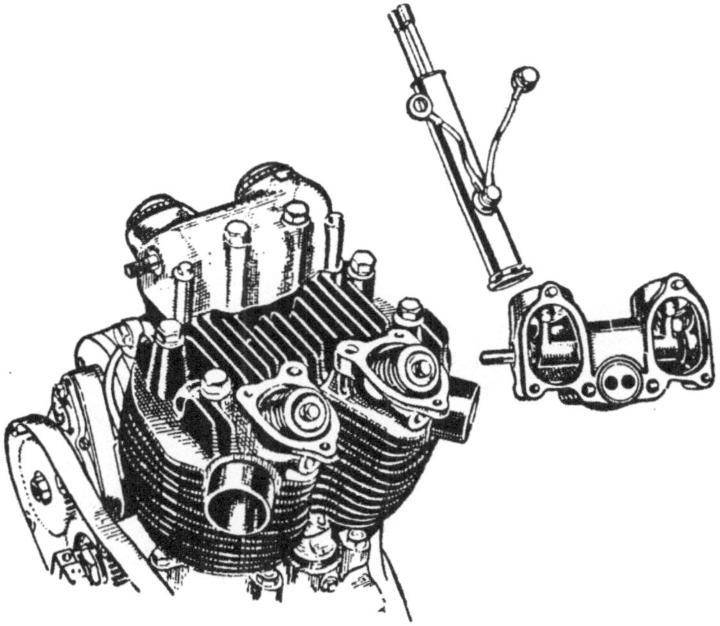

FIG. 19.—THE 1938/40 500-C.C. TWIN ENGINE WITH THE EXHAUST ROCKER-BOX REMOVED.

1938/40 and 1950/51 models are fitted with the oil drain pipes.

(5) Fit push-rod tube rubber washers over tappet blocks; inlet push-rods and tube should now be positioned to the tappet block, complete with drain pipes where fitted.

(6) Refit cylinder-head complete with manifold to cylinder block and lightly tighten the four outside bolts.

(7) Assemble inlet rocker-box to cylinder-head. Check for correct positioning of push-rods to rocker balls. Carry on the same procedure with the exhaust rocker-box and push-rods.

(8) The four centre cylinder-head bolts can now be fitted and the head evenly tightened down. Replace the four oil-drain adaptors where fitted.

(9) Adjust valve tappets (see data sheet).

(10) Replace carburetter, torque stays, sparking-plugs, H.T. cables, rocker oil-feed pipe and exhaust pipes.

(11) Refit petrol tank and saddle, connect up oil-pressure gauge and tighten down instrument panel.

(12) Run engine and carry out stationary tests. After stopping re-tighten all nuts and bolts.

Reassembly of TR5 Engine

(1) Carry out instructions on pages 56–57, paragraphs 1 to 5.

(2) Reassemble the rocker-boxes to the cylinder-head, but do not fit inspection caps.

(3) Fit the upper push-rods cover seal washers to the rocker-boxes and place the cylinder-head in position. Engage the inlet push-rods with the rocker ball pins.

(4) Raise the front of the cylinder-head and assemble the exhaust push-rods and cover tube. Carefully lower the cylinder-head to ensure correct assembly and inspect the cover tubes for seating over the tappet block and to the rocker-boxes.

(5) Fit the four short bolts and lightly tighten down the cylinder-head. Check valve operation, now fit the four remaining bolts and tighten down the head evenly. Replace the oil-drain bolts.

(6) Adjust valve tappets (see data sheet). Lubricate rocker gear through inspection holes and replace inspection caps.

(7) Replace induction manifold.

(8) Refit torque stays, carburetter, sparking plugs, H.T. cables, rocker oil-feed pipe and exhaust pipe.

(9) Run the engine and carry out stationary rests. After stopping re-tighten all nuts and bolts.

Reassembly of 3T Engine

(1) Replace pistons—ensure correct positioning of circlips and that the gaps in the piston-rings are not in line.

(2) Prepare cylinder block—grease base washer and fit to block. Fit the four tappets and secure against falling into the crankcase. Cylinder block can now be fitted to crankcase. The block must be installed with the "DS" mark, which will be found on the base flange, on the drive side of the engine.

(3) Fit rubber seal washers to push-rod tubes. Grease upper push-rod tube washers, cylinder-head gasket and fit the cylinder-head.

(4) Assemble inlet push-rods and tubes to cylinder-head. To aid assembly the lower push-rod tube cups are marked with a vertical line. This must be centralised and placed facing the

magneto and dynamo respectively. Fit the four centre holding-down bolts and torque bracket to cylinder-head and place in position on the cylinder block.

(5) To fit the exhaust push-rods, the cylinder-head should be raised about 1 in. This can be done by placing a piece of wood either side between the cylinder-head fin and block fin. Position both inlet and exhaust push-rods on the tappets, remove the wood pieces and gently lower cylinder-head to block. Inspect the rocker balls for correct assembly to push-rod cups. Now lightly tighten down the four centre bolts.

(6) Fit the four remaining bolts and evenly tighten down the cylinder-head.

(7) Adjust tappets and replace covers.

(8) Refit torque bracket to frame clip, the carburetter, sparking-plugs, H.T. cable, and the rocker oil-feed pipe and exhaust pipes.

(9) Run engine and carry out stationary tests. After stopping re-tighten all nuts and bolts.

REMOVING ENGINE FROM FRAME

All Twin Models

Before commencing work on the removal of the engine on all models except 3T, it is advisable to lift the tank in order to gain easy access to the engine (see page 54) and to drain the engine oil.

(1) Remove the exhaust system.

(2) Remove carburetter, leaving the slides attached to the cables, and tie the slides up in a safe position.

(3) Disconnect the plug leads, the magneto cut-out cable, the dynamo wires and ignition-control cable at handle-bar where fitted.

(4) Remove the torque stays. The 3T and TR5 models have a torque bracket bolted to the cylinder-head and connected to the frame by a clip and bolt. Remove the bolt and slide the clip along the frame tube.

(5) Disconnect all external oil pipes except rocker drain.

(6) Remove the foot-rests and foot-rest spindle.

(7) Remove the battery and battery carrier. 1949/51 models are fitted with an air filter ; this can be detached with the battery carrier.

(8) Take off the primary chaincase, clutch and engine sprocket (see page 91 for clutch removal).

(9) Slacken off the clip securing the dynamo to crankcase; remove the dome nut on the gear-case cover and withdraw dynamo from its housing. 1938/9/40 models will not require this operation, as the dynamo and magneto are one unit.

(10) Before commencing to remove the engine-plate bolts the engine should be supported under the crankcase with a box of suitable height. The rear engine-plate back bolts should now be removed and the rear-plate front bolts slackened off only. Remove all the front engine-plate bolts except the lower front bolt which connects rear frame to front frame; this should only have the nuts slackened off.

(11) Tilt the engine backwards and swing the front engine plates forward.

(12) Lift engine from frame.

FIG. 20.—THE SIMPLE JIG RECOMMENDED FOR HOLDING THE ENGINE ON THE BENCH.

DISMANTLING THE ENGINE, 5T, 6T, TR5 AND T.100

Having removed the engine from the frame, it should be mounted on the bench, so that work on it can be conveniently carried out. The illustration shows a useful and simple method of holding the engine firmly. Should the vice be of light construction, it is advantageous to fit a distance piece at the end opposite to that which is holding the crankcase lug. This arrangement has the advantage that there is no heavy jig required to hold the engine.

Dismantling Procedure

(1) Remove the oil-pressure release-valve, and take off the timing-gear cover after removing the securing screws.

(2) Unscrew the two nuts securing oil-pump, and slide the pump off the studs.

(3) To remove the magneto driving gear and auto advance device the instructions on " Magneto Removal " should be followed. The removal of the driving gear on models 1938/40 and the " Trophy " and 1951 T.100 is carried out by using the withdrawal tool (Part Number DA 50/1) supplied in the Standard tool-kit.

(4) Take off the magneto or magdyno as fitted.

(5) Remove the nuts from the mainshaft gear (right-hand thread) and the cam gear wheels (left-hand thread).

(6) Remove the timing gear. To withdraw the camwheels and mainshaft pinion, special tools are required. The camwheels are machined with a screwed boss to take the withdrawal tool (D178) and for post-war models a pinion extractor (ST1101W) is available. Early twin models: method of removing the latter is a little more difficult, operations 8 to 14 should first be caried out. The crankcase can then be supported on two metal blocks, gear side uppermost, and the assembly put under a press. When turning the press, pressure should be gently applied to force the crankshaft assembly through the pinion and ball-race.

(7) Take off the cylinder-head complete with rocker-boxes by unscrewing the eight holding-down bolts. The rocker-boxes can be detached after cylinder-head removal. 1938/40. and 1950/51 models—first remove the oil-drain pipes.

(8) Remove push-rods and cover tubes.

(9) Remove the cylinder block holding-down nuts and take off cylinder block. To secure the tappets against falling into the crankcase, a rubber band should be fitted over them before lifting the block.

(10) Remove the pistons by taking out a circlip in each piston and pushing out the gudgeon-pin. The gudgeon-pin can either be removed by using a removal tool or by tapping out with the use of a soft metal drift. The pistons should be suitably marked on the inside in order that they may be replaced in their original position.

(11) The remaining crankcase bolts should now be removed. Do not omit to remove the two screws fitted to the internal bosses.

(12) Take the engine out of the vice and remove the crankcase oil filter assembly.

(13) Split the crankcase by " bumping " the end of the mainshaft on a piece of soft wood. If the wood is placed on the floor and the bumping operation carried out in this position, the assembly will easily separate.

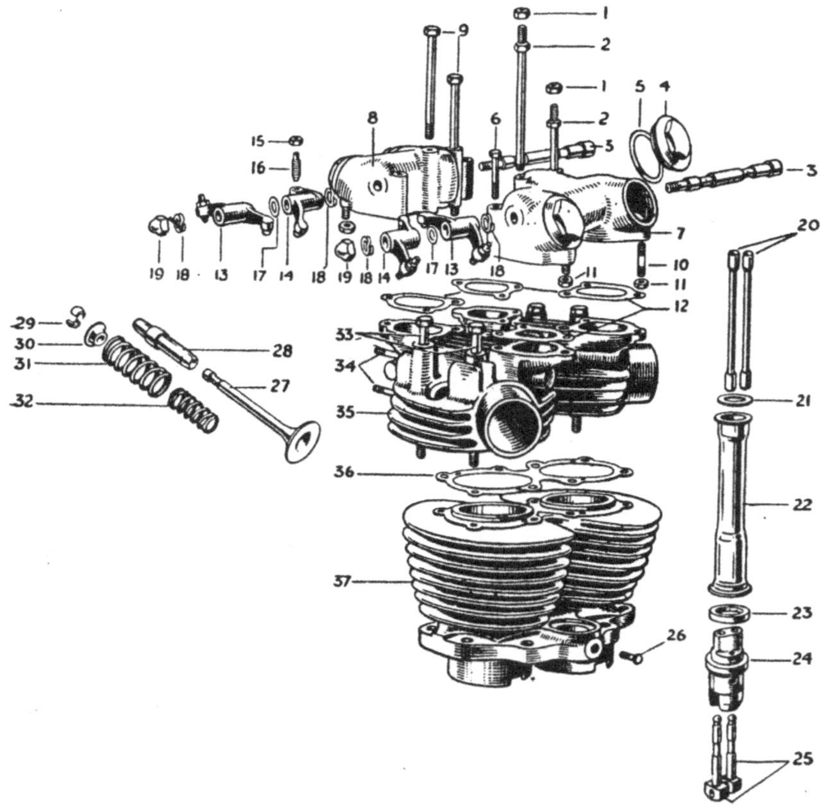

FIG. 21.—EXPLODED VIEW OF THE CYLINDER ASSEMBLY OF THE 500-C.C. POST-WAR TWIN ENGINE (1945-9).

The crankcase assembly is shown on the opposite page.

KEY TO PART NOS.

1. S1-52	12. E1650	22. E2273	30. E1543
2. E1596	13. E1458	23. E1496	31. E1487
3. E1512	14. E1457	24. E1477	32. E1488
4. E1564	15. E470	25. E1478	33. E327
5. E1577	16. E1513	26. H200	34. E1715 (T.100) E1554 (5T)
6. E1545	17. E1575	27. E1955 E1956	35. E2258 (5T) E2258A (T.100)
7. E1462	18. E1574	28. E1480 E1604	36. E2248
8. E1461	19. E1435	29. WE259A	37. E2257
9. E1484	20. E1693		
10. E1540	21. E1497		
11. F879			

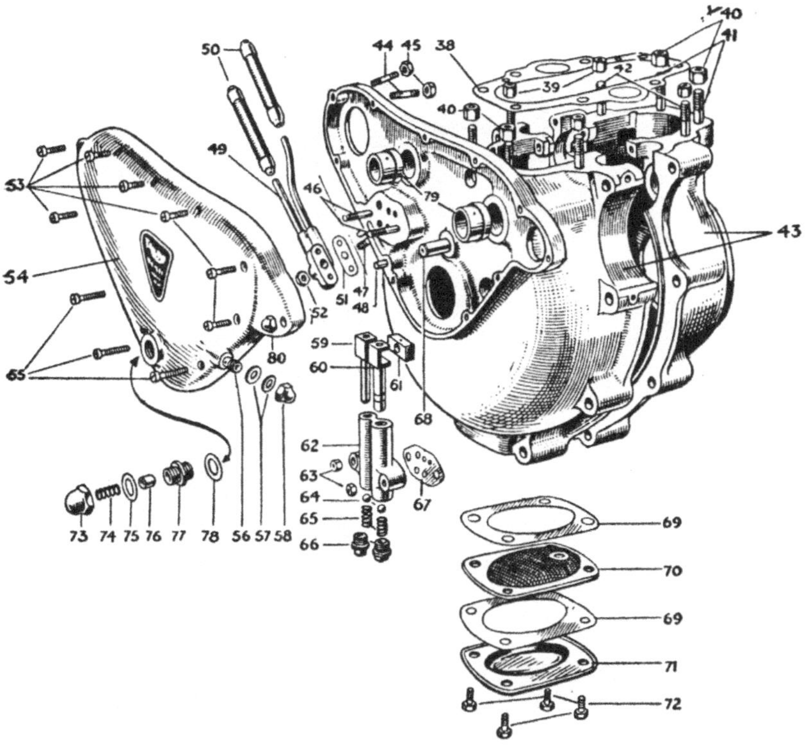

FIG. 22.—EXPLODED VIEW OF THE CRANKCASE ASSEMBLY OF THE 500-C.C. POST-WAR TWIN ENGINE (1945-9).

The cylinder assembly is shown on the opposite page.

KEY TO PART NOS.

38. E2249	49. E1958	59. E1475	70. E529
39. W76	49. E1526 (T.100, 5T)	60. E1476	71. E486
40. PO172A	50. MF119	61. E495	72. MT63
41. E672	51. E1529	62. E1470	73. E2304
42. E1614	52. S1–51	63. F879	74. GS250
43. E2219/20	53. E427	64. S70–7	75. GS254
44. E2022	54. E2218	65. E403	76. GS251
45. E2023	55. E1049	66. E386	77. GS248
46. E1536	56. E1586	67. E1537	78. E1670
47. E1539	57. E1335	68. E375	79. E1479
48. E1532	58. E1435	69. E487	80. E2228

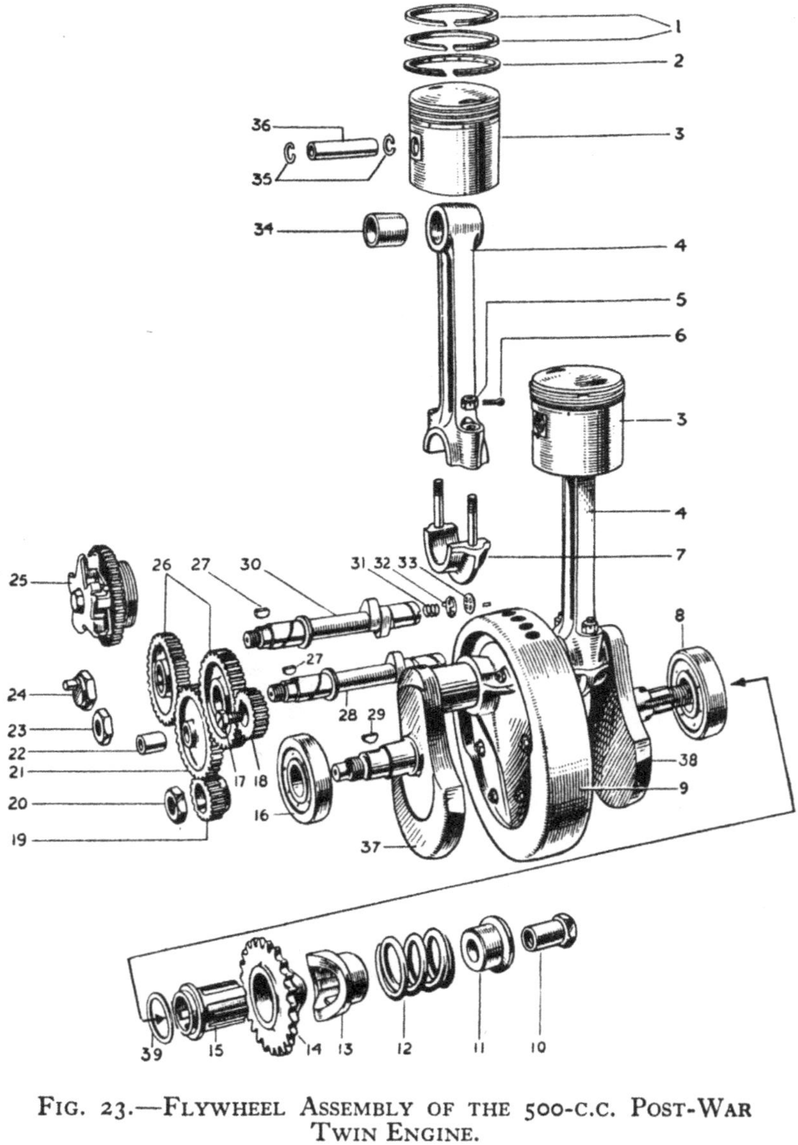

FIG. 23.—FLYWHEEL ASSEMBLY OF THE 500-C.C. POST-WAR TWIN ENGINE.

See page 65 for Part Nos.

KEY TO PART NOS. OF FIG. 23

1. E656	14. E2826C	26. E1486A
2. GE29	E2826F (sidecar)	27. E1558
3. E1563 (5T)	15. E1521	28. E2302
E1908 (T.100)	16. E2877	29. E1580
4. E2921	17. 111,700	30. E2302
5. E2922	18. E2451	31. E2256
*6. 1/16" × 3/4" split pin	19. E1472	32. E2254
7. Sold with E2921 only	20. E398	33. GS87
8. E1591	21. E1471	34. E1511
9. E1510A	22. E400	35. WE479
10. E509	23. E1492	36. WE203E
11. E396	24. E1463	37. E1459
12. E395	25. E1992 (B.T.H.)	38. E1460
13. E1681	47502A (Lucas)	

* Used with early pattern rod only

(14) Remove the crankshaft assembly and camshafts from the crankcase. When removing the inlet camshaft be careful not to loose the breather rotary valve and spring. (Post-war models only.)

(15) To remove the crankcase main bearings from the crankcase, gently warm the area around the bearings and drop the case from a height of about 6 in., open side downwards. Another method is to tap the bearings out with a suitable drift. Do not forget to remove the circlip before extracting the timing-side bearing.

Dismantling Crankshaft Assembly, 5T, 6T, TR5 and T100

Grip the lower part of the flywheel in the vice and take off the connecting-rods by removing the nuts and splitting the big-ends.

To dismantle the crankshaft assembly it is only necessary to take out the six nuts and bolts (these nuts and bolts should be discarded as they are not fit for further use). The driving-side and timing-side cranks can now be removed from the flywheel.

Before splitting the assembly, the flywheel and one of the crankshafts should be suitably marked in order to have the correct sequence when reassembling, otherwise the balance of the crankshaft will be affected.

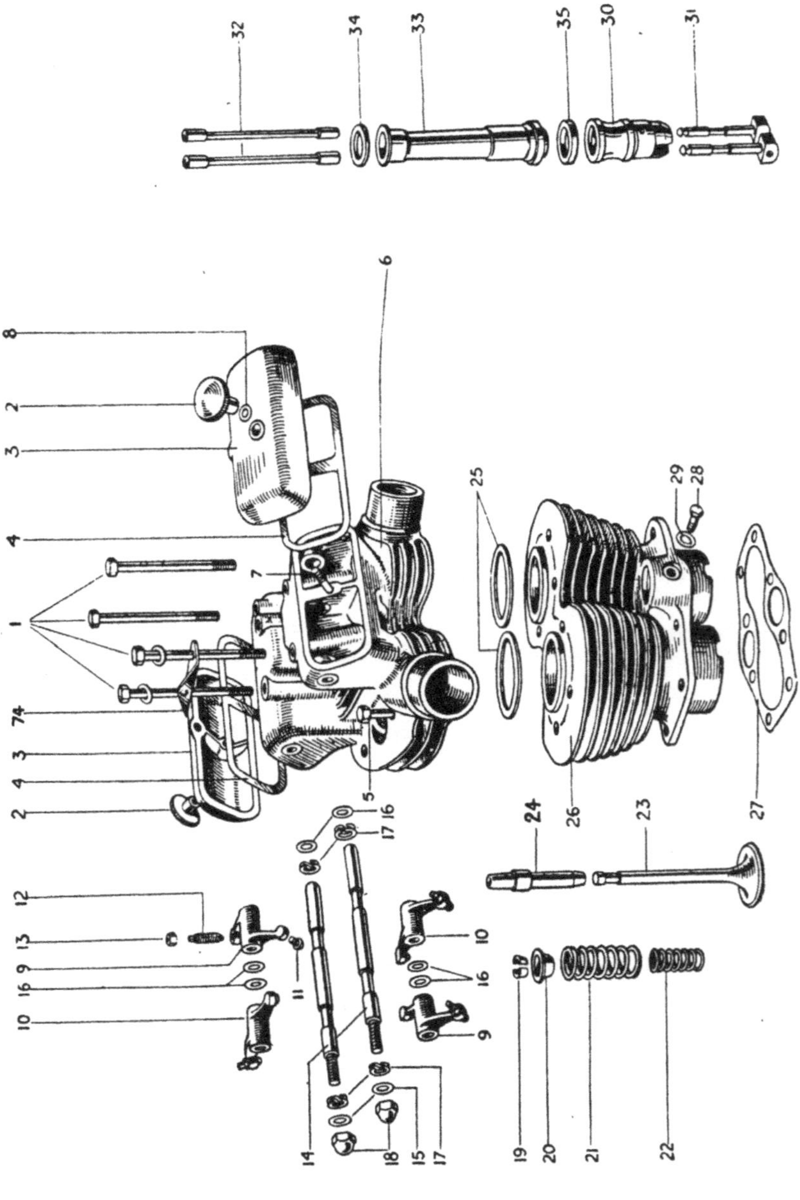
32
34
33
35
30
31
8
2
3
4
1
74
3
4
2
6
7
5
25
29
28
26
27
24
23
12
13
9
16
10
11
14
16
17
18
15
17
10
16
9
19
20
21
22

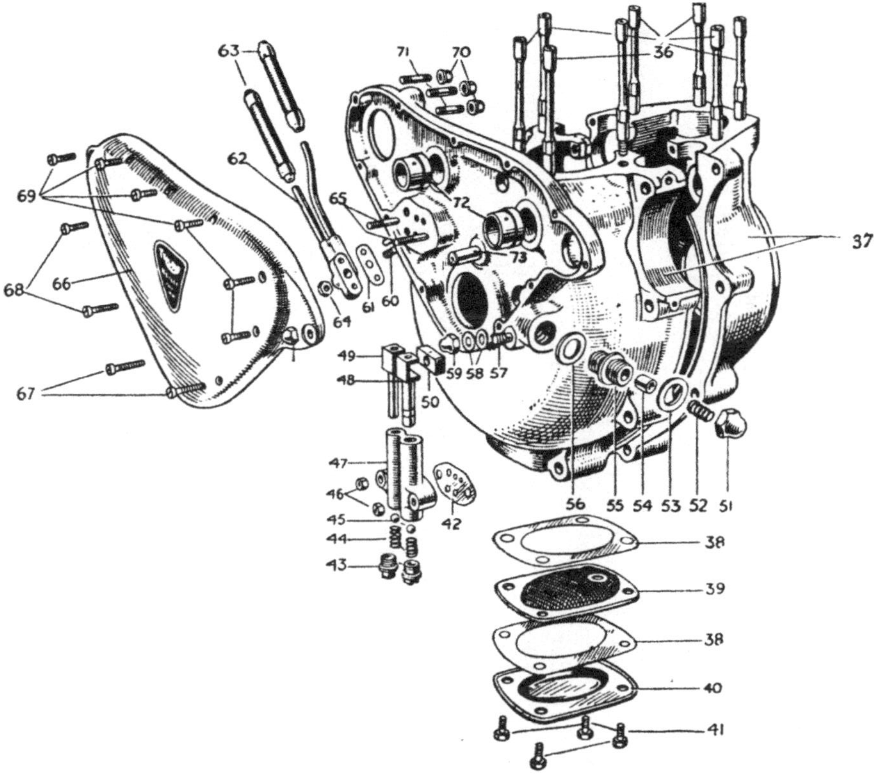

FIG. 24.—THE CYLINDER AND CRANKCASE ASSEMBLY OF THE 350-C.C. TWIN ENGINE (1ST CONDITION).

KEY TO PART NOS.

1.	E1795	38.	E487
2.	E1912	39.	E529
3.	E1779	40.	E486
4.	E1780	41.	MT63
5.	E1796	42.	E1537
6.	E2014	43.	E386
7.	E1987	44.	E403
8.	PT104B	45.	S70-7
9.	E1755	46.	F879
10.	E1756	47.	E1470
11.	E1483	48.	E1476
12.	E1513	49.	E1475
13.	E470	50.	E495
14.	E1813	51.	E2304
15.	E1330	52.	GS250
16.	E1575	53.	GS254
17.	E1574	54.	GS251
18.	E1435	55.	GS248
19.	WE259A	56.	E1670
20.	E1757	57.	E1586
21.	E1778	58.	E1335
22.	E1777	59.	E1435
23.	E1751 IN E1752 EX	60.	E1539
		61.	E1529
24.	E1759	62.	E1526
25.	E1776*	63.	MF119
26.	E1740*	64.	S1-51
27.	E1775	65.	E1536
28.	H200	66.	E2252
29.	E2312	67.	E1049
30.	E1837	68.	E427
31.	E1478	69.	E452
32.	E1785	70.	E2023
33.	E1838	71.	E2022
34.	E1497	72.	E1479
35.	E2425	73.	E375
36.	E1791*	74.	F2380
37.	E2250/51		

* See 1948/51 Spares Lists for modified Cylinder Block fixing.

Dismantling 3T

Mount the engine in the vice as described on page 59 and proceed with the dismantling as follows :—

(1) Remove the timing-gear cover and oil-pressure release-valve.

(2) Take off the oil-pump complete by removing the two nuts and sliding the pump off the studs.

(3) To remove the magneto driving gear and auto advance device, the instructions on "Magneto Removal" on page 150 should be followed.

(4) Take off the magneto by removing the three securing nuts.

(5) Remove the nuts from the mainshaft (right-hand thread) and cam gear wheels (left-hand thread).

(6) The cam gear wheels and the mainshaft gear pinion are removed from the shafts by the use of a special Triumph withdrawal tool which screws on to the threaded portion of the gear. The central bolt in the extractor is screwed in and the gear is pulled off the shaft.

(7) Unscrew the eight cylinder-head holding-down bolts and remove the head from the cylinder block.

(8) Remove the push-rods and cover tubes.

(9) Cylinder block can now be lifted, but before doing so the tappets should be secured against falling into the crankcase. To do this place a short ¼-in. bolt between the tappet grooves.

(10) Remove the pistons by taking out a gudgeon-pin clip from each piston and pushing the gudgeon-pins clear of the small-end bush. If a gudgeon-pin extractor tool is not available the gudgeon-pins may be tapped out with the use of a soft metal drift, but the opposite side of the piston must be well supported during this latter operation. Each piston should be suitably marked on the inside of the piston skirt to assist in the correct reassembly.

(11) Remove the remaining crankcase bolts, not forgetting the two clamping screws situated in the crankcase internal bosses.

(12) Take the engine out of the vice and remove the crankcase oil filter assembly at the bottom of the case.

(13) The crankcase is now ready for splitting; this operation is carried out by "bumping" the end of the drive shaft on a piece of soft wood. If the work bench is not sufficiently rigid this operation can be carried out more successfully by placing the wood on the floor. It will be found that the assembly will be easily separated.

(14) Remove the crankshaft and camshafts from the crankcase. When removing the inlet camshaft care should be taken not to lose the breather rotary valve disc and spring where fitted.

(15) To remove the crankcase drive-side main bearing from the crankcase, the crankcase should be gently heated round the bearing area to approximately 100° C. (about boiling-point). The crankcase half should now be dropped from the height of 6 in. open side downwards in order to remove the bearing from its housing. The removal of the timing-side plain bearing is not desirable if the fitter is not fully competent to insert a new white-metalled bearing and fit it to the crankshaft. See page 86 for instructions on refitting plain bearings.

Dismantling 3T Crankshaft Assembly

To remove the connecting-rods from the crankshaft assembly, it will be necessary to detach the cranks from the flywheel. In order to carry out this operation, first remove the two pinch bolts which attach the cranks to the flywheel, now place a soft metal drift through the aperture in the flywheel so that it contacts one of the cranks, then by giving the drift a few sharp blows with a hammer the crank and flywheel will be separated. Carry out the same operation to detach the opposite crank. Both connecting-rods can now be removed for examination. Between the two crankshafts an oil tube will be found; take great care not to damage this part, otherwise crankshaft oil pressure will be lost.

Note.—If the engine is removed due to failure of the white-metalled big-ends, it is essential that the crankshaft assembly is stripped and thoroughly cleaned in order to remove any foreign matter in the oilways. Parts such as oil-tank, filters, oil-pipes and oil ducts should be washed and blown out with high pressure air.

INSPECTION OF PARTS

Rocker-Box and Rocker-Gear

The 350-c.c. Twins have the rocker-boxes cast integral with the cylinder-head. To remove the rockers and spindles

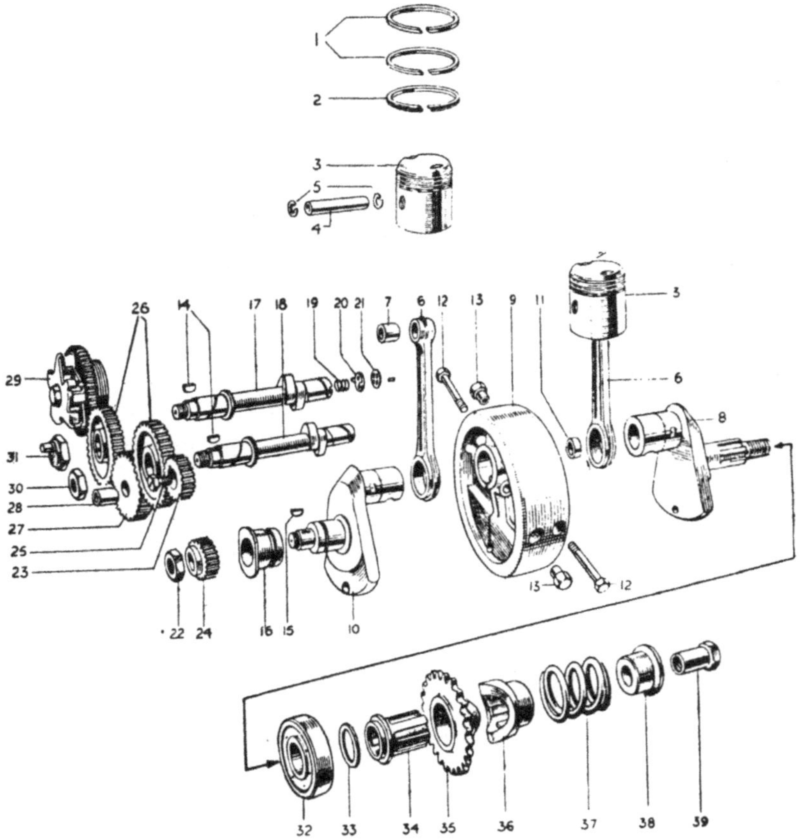

FIG. 25.—THE FLYWHEEL ASSEMBLY OF THE 350-C.C. TWIN ENGINE.

KEY TO PART NOS.

1. E1766	14. E1558	28. E400
2. E1819	15. E1580	29. E1992 (B.T.H.)
3. E1748	16. E2266	498152 (Lucas
* E1841	17. E2266	30. E1463
4. E1750	18. E1754	31. E1492
5. E1767	19. E2256	32. E1592
6. E1749	20. E2254	33. E1732
7. E1762	21. GS87	34. E1521
8. E2200	22. E398	35. E2826F
9. E1745	23. E2451	36. E1681
10. E1747	24. E1771	37. E659
11. E1790	25. 111,700	38. E396
12. E1760	26. E1486	39. E509
13. E1797	27. E2202	

* High compression.

it is only necessary to tap out the spindles from the off-side and extract the rockers from the rocker-box. The rocker can now be checked on the spindle for wear. Examine the ball pin which contacts the push-rod. A fair judgment of the amount of wear on this can be made by placing the push-rod to the ball pin at approximately its working angle. If the push-rod end contacts the rocker-arm, the ball pin should be changed. The adjuster pins should be examined for pit marks; if the marks are only slight they may be removed with the use of an oil-stone. With the other machines, the rocker-boxes are detachable from the cylinder-head. Method of stripping is identically the same as the 350-c.c. models.

Cylinder-Head (Light Alloy)

This type of cylinder-head is fitted to the TR5 and T100 only, and when removing the carbon great care should be taken not to mutilate the head sphere and ports. A round spade-type tool is suitable for removing the heavy carbon deposit, and the final finishing can be carried out with smooth emery cloth. Whilst doing this work take care not to damage the valve seatings. Wash the head thoroughly in a cleaning solvent after decarbonising. The exhaust-pipe stubs on the T100 are screwed into the exhaust ports, and you are advised not to remove them, but at all times ensure they are tight. *Do not use a caustic-soda bath when servicing the alloy head.*

Cylinder-Head (Cast Iron)

Where convenient it is desirable to boil the cylinder-head in a solution of caustic soda, which will effectively remove all carbon from inaccessible positions. Do not forget to remove the alloy induction manifold from the cylinder-heads, as caustic soda has a most detrimental effect on aluminium. As this process removes the paint from the cylinder-head, the head should be copiously black leaded or painted with a good heat-resisting enamel before refitting. The T100 cylinder-head should be polished in the same manner as the alloy type. Great care should be taken not to damage the valve seatings, otherwise this will mean recutting the seatings, a practice which should only be carried out when new valve-guides have been fitted, or after considerable mileage.

Valve-Springs

These parts should be thoroughly inspected, exhaust springs particularly, for any signs of cracks and fatigue. It is essential that the springs are in good condition, otherwise the performance of the machine will be affected and if there is any doubt, a new set of springs should be fitted. The normal life of a valve-spring is 10,000 miles.

Valve Grinding

This operation should be carried out in exactly the same manner as described on page 18.

Piston

If the cylinder block is not removed during decarbonising it will only be necessary to clean off the carbon on top of the piston. To avoid scratching the piston, it is a good plan to obtain a stick of solder and hammer it out flat at the end. It is advisable not to disturb the carbon around the top of the cylinder bore, and about $\frac{1}{16}$ in. around the rim of the piston, as the carbon at this point forms an oil seal.

Piston Inspection

When the pistons are removed for inspection or the fitting of new piston-rings, they should be thoroughly cleaned, but care should be taken to avoid removing any metal from the piston lands. Remove all the carbon from the piston-ring grooves and test for depth, clearance and wear by rolling a piston-ring around the groove. Test the gudgeon-pin piston bosses for wear by placing the gudgeon-pin in position; this should be a good tight fit, as any slackness here will result in a mechanically noisy engine. The pin should be a floating fit in the small-end bush.

Piston-Rings

When it becomes desirable to fit new piston-rings they should first be checked for depth by placing the outer edge of the ring into the piston-ring groove. The ring should roll freely around the groove; if the clearance is excessive the ring will have a pumping action, thereby causing a heavy oil consumption. A tightly fitted piston-ring will either gum up badly or cause a piston seizure. The piston-ring gap should be 0·010 in. to 0·014 in. when new. To check this, place the ring in the lower part of the cylinder bore and square up by placing the piston on top of the ring. Never check the piston-ring gap anywhere in the working area of the cylinder bore.

Cylinder Block

Wash the cylinder block thoroughly, and well dry out the bores afterwards with a piece of clean cloth. Hold the block up to the light and inspect for score marks and any sign of aluminium adhering to the cylinder walls due to seizure. If any trace of aluminium is found in the cast iron block, this can either be removed by boiling the block in a strong caustic soda solution or by carefully removing the aluminium away with a bearing scraper. With the light-alloy block only the latter process *must* be used.

Checking Wear of Cylinder Bore

To check the amount of wear which has taken place in the bores, the carbon deposited around the rim of the bore should first be removed; an internal micrometer or preferably a Mercer gauge can then be employed to estimate the wear. If a reading is taken in the lower portion of the bore and another at the top, the difference between the two readings will indicate the amount of wear which has taken place.

Boring Out Cylinder Block

The cast-iron cylinder block can be bored out to a maximum of 0·040 in. In no circumstances should this boring figure be exceeded, as the walls of the cylinder will be thinned too greatly, causing over-heating and probably seizure of the pistons. For the same reasons sleeving is not advised. The light-alloy block, as fitted to the 1949/50 TR5 model, has the liners cast into it, and they cannot be removed, but as the liner is of sufficient thickness, maximum boring size is also 0·040 in.

The 1951 TR5 and T100 has a completely re-designed die-cast light-alloy cylinder block, into which special cast-iron liners are fitted. The maximum re-bore for this type of block is 0·020 in. No attempt should be made to replace the liners after wear has taken place beyond this stage; the block should be replaced. Although the re-boring figure is less than with the cast-iron block, the mileage will be far greater before this operation becomes necessary.

Tappet Guides

It is not necessary to remove the tappet guides from the cylinder block for inspection; the amount of wear can be estimated by rocking the tappet head. To remove the tappet guides, place the cylinder on the bench, remove the lock-screws securing the guides. Then by placing a soft metal drift between the guide lips, the guide can be tapped from its housing.

Tappets

As these parts are constantly submerged in oil, wear is slight. Small pit marks on the base of the tappet can be removed with the use of an oil-stone. If the marks are too deep it will, of course, be necessary to fit new parts, as any great amount of metal removed from the tappet head will weaken the structure and remove the case hardening.

Push-Rods

First examine the push-rods for straightness; to do this lay the push-rod down on a true surface (a piece of plate glass can be used in the absence of a surface plate) and roll the push-rod over until its full circumference has been covered. Any irregularity will easily be seen. Now examine the push-rod ends, for fracture or slackness on the tube, and renew if any failure is noticed.

Push-Rod Cover Tubes

If oil leakage is to be avoided after overhaul make sure the cups on the cover tubes are tight and in no way damaged. Any slackness of the cups can be overcome by resoldering the parts together.

Crankcase

The crankcase should be thoroughly washed and then inspected for cracks. If there is any doubt about a crack, the chalk test should be applied. Place the crankcase halves into hot oil (100° C.), leave the parts submerged for about 10 minutes, then take them out and thoroughly dry. When the crankcase is perfectly dry apply chalk dust all over. Immediately the crankcase starts to contract the oil will be squeezed out of the suspected crack and discolour the chalk, thereby making easy recognition. If the camshaft bushes prove to be worn, the timing-side bushes can be pressed out, but it will be necessary to tap a thread in the driving-side bushes. The tap should be left in position and the squared end gripped in the vice; now gently tap the crankcase away from the bush until the two parts are separated. If any difficulty is encountered, remove the tap from the bush and apply heat locally around the bush, replace the tap and carry out the same procedure as before. When the new bushes have been replaced it will be necessary to bolt the two halves of the crankcase together before carrying out the reaming operation. It is most essential that these bushes are line-reamered.

Timing Cover

The timing cover on the 350-c.c. models will require no attention other than a check on the jointing face before replacing. With the 500-c.c. and 650-c.c. models, however, the timing-case cover supports the crankshaft sealing bush; if the bush shows signs of wear it must be replaced, otherwise lack of oil pressure to the big-end assembly will result. To remove the bush a thread should be tapped in it. Leave the tap in position and place the squared end in the vice; the cover can then be tapped off the bush. On the models after 1940 the bush can be pressed out and the new one replaced in a like manner. If this work is to be carried out whilst the engine is still in the frame, it will be necessary to remove the O/S exhaust pipe and turn the cover through 180 degrees; the fit of the bush can now be tested on the crankshaft.

Timing Gear

Timing gear wear is negligible; the only part that needs changing after considerable mileage is the intermediate gearwheel bush. After the old bush has been removed the new one should then be pressed into the gear wheel and reamered in position. The spindle on which this gear operates should be checked for security in the crankcase; if found to be loose, the part of the spindle that enters the case should be tinned and replaced. Take great care when dismantling the magneto

FIG. 26.—THE TIMING GEAR AND OIL-PUMP OF THE 350-C.C. TWIN ENGINE.

The assembly on the post-war 500-c.c. twin engine is similar.

gear; any damage caused to the auto-advance mechanism will affect the performance of the machine (see page 162 for magneto gear wheel removal).

Camshaft Bushes

To check the amount of wear in the camshaft bushes it will be necessary to remove the cylinder-head or release the tappet adjustment in order to relieve all pressure off the camshaft. Now remove the timing cover and detach the intermediate gear wheel from its spindle. The camshaft gear can now be gripped and the amount of wear estimated by lifting the gear.

Wear in camshaft bushes is practically negligible and it will only be necessary to change them after considerable mileage.

Camshaft

The camshaft will require no servicing other than cleaning. It will be noted that in the 1945 models and onwards the inlet camshaft houses the breather rotary valve. The aperture into which the valve fits should be well cleaned and the spring checked for freedom of movement. If the cams show any sign of wear the camshaft must be discarded, as the relieving of any score marks or pitting will alter the contour of the cam and seriously affect the performance of the motor cycle.

Oil-Pump

Wash the pump thoroughly in petrol and allow to dry before attempting to inspect the parts. First, inspect the aluminium driving block for any signs of ovality in the centre hole; this test can be carried out by placing the block on its driving shaft. Should the block be worn it must be discarded, as the stroke of the plunger will be affected, causing a reduction in oil pressure and inefficient scavenge. The plungers and oil-pump body will show very little wear as these parts are constantly running in oil. Inspect the ball valve seatings for pitting; if the marks are on the surface the balls should be replaced and given a sharp tap on to their seatings. If the pit marks are too deep to be removed in this way, the seatings should be re-cut, but only a minute amount of metal should be removed. Assemble the oil-pump and check its action by submerging in oil and pumping the plungers.

Big-End Bearings

These can be examined for wear after the cylinder block has been removed. With this type of connecting-rod a considerable amount of end float is necessary. To test the up-

and-down movement of the connecting-rod on its journal, turn the crankshaft over until the connecting-rod comes up to top dead centre, steady the rod with one hand and lift the rod with the other; the amount of play should be barely perceptible.

Crankshaft

If the motor cycle has been in operation for a considerable mileage it is advisable to split the crankshaft (see page 65 for 500-c.c. and 650-c.c. and page 68 for 350-c.c. models) as a certain amount of oil sludge will be found in the internal oilways, although with modern detergent oils most of this sludge is kept in suspension and there is little likelihood of blocked oilways. To remove this oil sludge the crankshaft can either be boiled in a strong solution of caustic soda or cleaned by scraping and then washing out in paraffin or petrol. When clean, the journals should be checked for ovality and score marks (see data sheet). If the journals are oval beyond the limit this will necessitate regrinding and the fitting of suitable undersize connecting-rods. Connecting-rods are supplied in the following recommended sizes: —0·005 in. and —0·010 in. Moderate score marks can be eased down by the use of fine emery cloth, which of course will not entirely remove them, but will assist in removing any rough edges. Next check the crank web mainshaft and gearshaft by positioning the ball races; these should be a push fit on the shafts. Any slackness at this point will cause a rough engine and rapid wear to the bearings. It will be noted that the 350-c.c. model has a white metal timing-side bearing and it is essential that the shaft is a good fit, as oil loss here will reduce the pressure of oil to the big-ends.

Main Bearings

Wash the bearings in petrol and allow to dry. The amount of play will now easily be noted; this should be only just perceptible. Main bearing wear can of course be tested before breaking down the crankcase by removing the primary chaincase cover and gripping the shock absorber firmly. By lifting the shock absorber the amount of wear in the bearings will be felt. To test the timing side, remove the timing-case cover and place a screwdriver under the half-time pinion and by levering up the amount of play will be noted. If any perceptible lift is felt in the last two tests the bearings must be replaced. The replacement bearing on the timing side should be the roller type as fitted to the later machines. For timing-side white-metallied bearing on 350-c.c. models, see page 86.

Crankcase Breather

The breather is of the rotary type and is incorporated with the inlet camshaft. The camshaft is hollowed out and slotted at the end to accommodate the spring and rotary valve. The valve disc is located at the back of the camshaft bush and is pegged to avoid rotation. On starting the engine the rotary valve will rotate with the camshaft and the crankcase pressure will exhaust through the ports of the valve and out of the breather pipe. Maintenance here is "nil" and it need only be checked when dismantling the engine.

The breather fitted to the 1938/40 models is of the diaphragm type and will be found screwed into the drive side of the crankcase. The diaphragm is located in the body of the breather, and on the downward stroke of the piston the diaphragm is forced off its seating, thus allowing the crankcase pressure to be released. On the upward stroke of the piston the diaphragm is sucked back on to its seating, thus sealing the crankcase from atmosphere. If the breather is exhausting too much oil and the scavenge side of the oil system has been checked and found correct, the breather should be removed and washed in petrol. This will dislodge any foreign matter which may have held the diaphragm off its seating. To test the breather, place diaphragm end in mouth and suck. If air can be drawn the assembly should be discarded and a new one fitted, as this would denote that the diaphragm and seating are faulty.

Oil and Gas Leakages

If correct servicing is carried out at and between decarbonising periods, oil leakages will not occur. When any joints are disturbed new gaskets should be fitted, and it is essential that these are of Triumph manufacture. Local-made gaskets are not usually conducive to an oil-free exterior of the motor cycle. Parts that are sealed with a jointing compound should have the faces well cleaned before applying a new coating. Take care not to damage the jointing faces when cleaning.

Gasket sets for decarbonising can either be purchased from your dealer or direct from the Triumph Spares Department. These sets will cover all washers and joints required except the primary chaincase gasket and cylinder-head gasket. The chaincase gasket is supplied separately, as this requires careful packing due to its size and material. The cylinder-head gasket need only be changed after considerable mileage has been covered. During decarbonisation of the engine, however, it is advisable to anneal this part by heating it over a gas burner to a blood red, then plunging it into a bucket of cold water. The gasket will now be quite soft, therefore allowing it to conform to any irregularity between cylinder-head and block.

All washers and gaskets, other than rubber ones, should be smeared with grease on both sides before assembly, as this assists in positioning the parts during erection and helps to make a gas- and oil-tight joint. There is, however, one exception to this procedure, that being the induction manifold washers between the cast-iron cylinder and aluminium manifold. Here it is desirable to coat the washers with a jointing compound to prevent them moving between the two dissimilar metals and covering up the induction aperture.

Assembly of Crankshaft Unit 5T, 6T, TR5 and T100

After ensuring that the cranks are perfectly clean and all oil sludge has been removed from the oilways, the crankshaft assembly can be built up. Assemble the cranks to the correct sides of the flywheel, as marked during dismantling. When fitting the crankshaft nuts and bolts, care should be taken not to overstrain them, as undue strain will cause stretch and may result in fracture of the bolt under load. Lock the nut to the bolt by centre punching between their threads.

Assemble the connecting-rods to the crankshaft journals in a dry condition; in this state the amount of lateral and vertical movement can be more easily detected. The lateral movement of the connecting-rod should be 0·025 in. to 0·030 in.; the vertical movement should not be perceptible, but freedom of the rod on the journal in all positions must be ascertained. If the vertical movement seems excessive and the white metal in the big-end cap is in good condition it is permissible to remove a skimming of metal off the alloy rod abutment face and to rebed the bearing. If the operator is at all doubtful about carrying out this last operation, he is strongly advised to return the assembly to the Works or his dealer for reconditioning. When finally assembling the connecting-rods, lubricate the journal with engine oil before fitting the rods, and the crankshaft should also be primed with oil by using an oil-pressure gun. As in fitting the crankshaft bolts, it is extremely important not to overstrain the connecting-rod cap studs or bolts when tightening up the nuts. The approximate amount of pressure on the spanner should be 18 ft.-lbs.

Reassembly of the 5T, 6T, TR5 and T100 Engine

(1) Crankcase.—Ensure that the two halves are perfectly clean and the old jointing compound has been removed from the surface faces.

(2) Fit the main bearings to the crankcase, not forgetting the bearing circlip securing the timing-side bearing.

(3) After lubricating the bearings, fit the crankshaft assembly to the drive-side half of the crankcase and insert the camshafts to their respective bushes. On the post-war model the inlet camshaft has the rotary valve fitted. (Rotary valve assembly is, first, the spring into the camshaft aperture, and secondly, the valve disc, which fits into the camshaft slots.)

(4) Smear jointing compound on the joint face of the timing-side crankcase and assemble to drive side. Bolt up the two halves and insert and screw up the two internal boss screws.

(5) Fit the crankcase to vice as in dismantling procedure.

(6) Fit the pistons to their respective connecting-rods and ensure the circlips are correctly positioned in the piston grooves.

(7) Prepare the cylinder block for fitting by greasing the paper base washer and fitting to block. Now lubricate and fit tappets to the tappet guides. Check the action of the tappets by working them up and down in the guides (this should be free and smooth) and then secure them against falling into the crankcase when the block is assembled. Lubricate cylinder bores.

(8) Where possible it is advisable to fit piston-ring clips to the pistons to aid assembly of the cylinder-block. Fit the block to crankcase and refit and tighten down the cylinder-base nuts.

(9) The valve-timing gears can now be fitted to their respective shafts. Fit the woodruff keys to the shafts and then assemble the crankshaft pinion to the crankshaft. This can be tapped into position with the use of a soft metal hollow drift. Now turn the engine over until the dot centre-punched on the gear is in the upright position. The assembly of the cam gears should be carried out with the Triumph assembler tool (D181). If the cam gears are tapped into position with a drift there is a possibility of damaging the camshaft bush, as the woodruff key will come into contact with the bush. *Note.*—Three keyways are cut in the cam gears fitted to the 1951 model, but only the one that is marked should be employed for standard valve timing. After the cam wheels have been fitted the intermediate wheel should be positioned on its spindle in such a way that the dots on the pinion and cam gear coincide with the dots on the intermediate wheel (see Fig. 27). Do not fit gear-retaining nuts yet.

(10) Grease and fit the cylinder head gasket to the cylinder block, assemble the head to the block, and lightly tighten down the four short bolts.

(11) Fit the lower push-rod cover oil seals (rubber) to the tappet guides, and assemble the push-rods with cover tube to the inlet side first.

(12) Grease the rocker-box gaskets and upper push-rod tube

FIG. 27.—A PHOTOGRAPHIC DIAGRAM OF THE TIMING GEAR
1951 models have three keyways cut in the cam wheels.

cover washer and place in position on the inlet rocker-box. Turn the engine over until both inlet push-rods are rocking, this puts them in their lower position for the assembly of the rocker-box. Fit the rocker-box, and after tightening down check the operation of the valve gear. For the exhaust side carry out the same procedure as above and paragraph (11). Fit remaining cylinder-head bolts and tighten head to cylinder block. Where oil-drain pipes are used, these should now be fitted.

(13) Fit the nuts to the cam gears and timing shaft pinion and tighten up.

(14) Prime the oil-pump with oil, ensuring that the plungers and sliding block are quite free, and assemble to engine. See that the paper washer installed between pump and crankcase is fitted so that all oilways are clear.

(15) Assemble the magneto to the engine and before finally tightening up the magneto securing nuts, fit the automatic advance device on magneto shaft but do not tighten up magneto gear unit. Where a magdyno is fitted the unit should be positioned on the platform. Screw up the set screw under the platform into the magdyno base; do not tighten. This should be tightened with the strap bolt after the magneto has been timed in order to get the correct pressure on the cork washer (see page 42).

(16) Time the magneto to engine (see page 163).

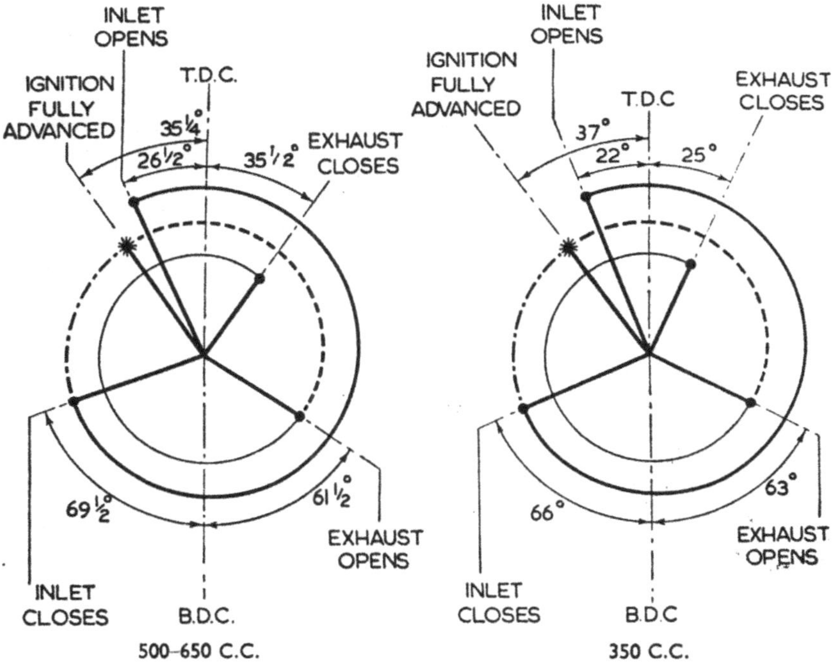

FIG. 28.—TIMING CHARTS.

(17) Smear jointing compound on to the timing cover joint face and replace cover to engine.

(18) Fit the oil-pressure release-valve to the timing-case cover.

(19) Adjust the valve tappets and replace inspection covers. Refit sparking-plugs.

(20) Remove engine assembly from the vice and pour about half a cupful of oil through the oil filter aperture, and assemble filter to crankcase.

(21) Lightly bolt the rear engine plates into position and the engine is ready for installation into the frame.

Reassembly of the 3T Crankcase

When refitting the main bearings, the area around the crankcase bearing housings should be heated to 100° C. (approximately boiling-point). This procedure ensures a more accurate assembly of the main bearings, and obviates any damage to the bearing housing, due to the expansion of the crankcase. When the timing-side crankcase has cooled down, the plain main bearing should be tested for any ovality. This may occur due to the structure of the crankcase. Where possible this bearing should be broached; if a broaching tool is not available it will of course be necessary to relieve any high spots by scraping away the surplus metal.

Carefully insert the crankshaft to the white-metalled bearing and rotate; the markings on the white metal will then plainly indicate the high spots. It will perhaps be necessary to carry out this operation two or three times before a 100 per cent. bearing surface is made. The crankshaft when finally fitted should rotate freely without any perceptible up-and-down movement. If the lubrication diagram on page 46 is studied it will be seen that the oil pressure to the crankshaft relies entirely on the fitting of this bearing.

Assembly of Crankshaft Unit 3T

Thoroughly wash the crankshaft big-ends and the adjacent hole in the flywheel with petrol to ensure against any possibility of crankshaft slip. The crank to the flywheel is an interference fit with tolerance of 0·001 in. minimum, 0·002 in. maximum. Now oil the crank surface upon which the connecting-rod fits, being careful to avoid the part that enters the flywheel. Assemble the connecting-rods to the cranks. On the connecting-rod will be found the marking " Flywheel Side ". This marking must face towards the flywheel. Fit the cranks to the flywheel and do not omit the oil tube that fits between the two cranks. Align the assembly by inserting a round bar ($\frac{1}{2}$ in. $-$ 0·003 in. $+$ 0·000 in.) through the three datum holes drilled in the cranks and flywheel. When the assembly is correctly aligned the cranks should be pressed into the flywheel. For this operation a hand-press should be used. Now that the assembly has been pressed together, it will be necessary to check the amount of side-play of the connecting-rods. Each connecting-rod should have between 0·015 in. and 0·020 in. lateral movement. Where any discrepancy occurs the crank on which the connecting-rod having the greatest movement is mounted should be pressed into the flywheel a little farther, until the movement of the connecting-rods shows an equal figure. This operation can be carried out by placing the flywheel on two blocks with the offending crank in the topmost position; the upper crank can now be pressed in the desired amount.

Before retightening the pinch bolts, check the alignment of the datum holes. If the assembly is not correctly aligned, tap the offending crank smartly with a lead hammer in the direction required, until the rod will slide through the holes easily. Tighten up the pinch bolts and recheck alignment. Two ball races of suitable size should now be fitted to the crankshafts and the assembly placed on a rig with only the ball races in contact. Rotate the crankshaft assembly and at the same time place a dial test indicator alternately on the drive shaft, timing shaft and flywheel rim. This will give a true reading of alignment from its axis.

Reassembling the 3T Engine

(1) Prepare the crankcase and oil all bearing surfaces liberally.

(2) Fit the crankshaft assembly to the drive-side crankcase. Now install the camshafts to their respective bushes. The camshaft with the breather rotary valve should be fitted to the rear bush (inlet).

(3) Apply a thin coat of jointing compound to the inner face of the timing crankcase and assemble the two crankcase halves. Bolt up the assembly and fit the two screws to the crankcase internal bosses.

(4) Now place the crankcase in the vice as shown in the illustration on page 60.

(5) Replace the pistons on their respective connecting-rods.

It is advisable to fit the gudgeon-pin circlips to the inner side of the pistons first to aid assembly. Fit the remaining circlips and ensure that they are correctly in their grooves.

(6) Lubricate and fit the tappets to their guides in the cylinder block, and check for smooth operation. Secure the tappets against falling into the crankcase by placing ¼ in. bolts between the tappet grooves.

(7) Grease the cylinder-base washer and position on the crankcase. Lubricate the cylinder bores and pistons with engine oil and assemble the cylinder block to the crankcase, with the mark D.S. stamped on the base flange to the drive side of the crankcase. To aid the assembly of the cylinder block piston-ring clips should be used.

(8) The timing gears should now be assembled. First position the woodruff keys to the camshafts and crankshaft. Fit the crankshaft pinion with the threaded portion away from the engine and tap in position by using a soft hollow drift. Assemble the cam-wheels to the camshafts by using the Triumph special cam-wheel press-on tool (D181).

(9) Insert the intermediate gear so that the punched marks on the gear coincide with the marks punched on the cam-wheels and pinion gears (see illustration on page 83). When the engine is turned one revolution it will be found that the timing gear marks do not coincide; this is due to the uneven number of teeth on the gears. The valve timing will of course be unaffected. Do not tighten the cam and pinion gear nuts until the upper half of the engine has been erected.

(10) Fit the rubber oil-seal washers to the lower end of the push-rod cover tubes. Locate the push-rods inside the cover tubes.

(11) Grease the cylinder-head gasket and upper push-rod cover-tube washers and fit to the head. Assemble the inlet push-rods and tube to the cylinder-head with the line marked on the cover-tube cup towards the rear of the machine.

(12) Fit the cylinder-head and push-rods assembly to the block. Lift the cylinder-head at the front and fit the exhaust push-rods and cover tube. Now examine the push-rods for location on the rocker balls. Fit the four small holding-down bolts and then check the action of the valve gear. Fit the four remaining central bolts with the torque bracket. It is essential that the final tightening of the cylinder bolts is carried out evenly.

(13) Tighten up the cam-wheels and pinion nuts.

(14) Assemble the oil-pump to the engine. See that the paper joint washer is correctly fitted.

(15) Fit the magneto and auto advance gear and time magneto to engine (see page 162).

(16) Smear jointing compound on the timing-gear cover face and fit the cover to crankcase. Position and tighten up oil-pressure relief valve.

(17) Adjust the valve tappets and replace inspection covers. Clearance when cold, 0·001 in.

(18) Remove the engine from the vice and fit the crankcase oil filter assembly.

(19) Fit the rear engine plates, but do not tighten up the stud nuts; this will help in assembly of engine to frame.

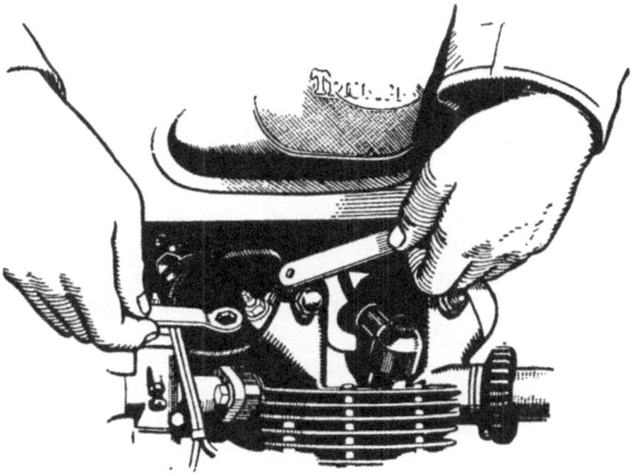

FIG. 29.—THE CORRECT TOOLS FOR TAPPET ADJUSTMENT ON THE SINGLE AND TWIN ENGINES WHERE THE ADJUSTING IS DONE BY LIFTING THE ROCKER-BOX CAP.

Installation of Engine to Frame, All Models

(1) Fit the foot-rest distance piece between the two rear engine plates, and support it in position by placing a long bolt through the foot-rest spindle holes in the plates.

(2) Place a box under the frame cradle as in dismantling and lift the engine into position.

(3) Tilt the engine backwards and at the same time swing the two front engine plates in position.

(4) Allow the engine to re-position itself and then fit the remaining crankcase and frame studs and nuts and tighten up.

(5) Fit the torque stays to the engine and tighten up the nuts. The 3T and TR5 engines will already have the torque bracket in position, therefore it will only be necessary to slide the torque clip on the frame bottom rail into position, and to fit and tighten up the nut and bolt.

(6) Refit the carburetter.

(7) Install the dynamo in its housing (post-war models) and connect up the cable wires, also magneto cut-out wire and control cable where fitted.

(8) Position foot-rest spindle and fit distance piece which is situated between nearside engine plate and primary case.

(9) Smear jointing cement in the inner primary case crank-case front aperture and fit the case in position. Ensure the oil seal is fitted under the spring plate and over the gearbox main-shaft. Fit and tighten up the bolt securing rear upper chain guard to chaincase.

(10) Fit the key to the gearbox mainshaft and the sleeve to the engine shaft. The clutch assembly less plates can then be fitted with the primary chain and engine sprocket in one operation.

(11) Select bottom gear and assemble the nut and washer to the gearbox mainshaft and screw up until the pressure on the nut turns the rear wheel.

(12) Now fit the shock absorber slider to the engine sprocket, followed by the spring, collar and finally fit and tighten up the engine-shaft nut.

(13) The nut securing the clutch housing can now be finally tightened and the lock washer turned over.

(14) Assemble the clutch plates to housing (see page 93).

(15) Fit the primary outer chaincase. Use a new paper gasket when fitting.

(16) Fit the brake pedal and foot-rests to their respective spindles.

(17) Replace battery and carrier and air filter where fitted.

(18) The oil block should now be fitted. Before doing so, however, it is advisable to have a quantity of oil in the oil-tank to ensure against an air lock in the feed-pipe.

(19) Fit the oil-gauge pipe and rocker-feed pipe to the engine.

(20) Clip exhaust pipes and silencers to engine and frame.

(21) Re-fit the sparking-plugs and H.T. leads.

(22) Lower tank in position and bolt to frame; ensure the rubber washers are in position. Connect up petrol pipes.

(23) Fit the saddle-nose bolt or replace the "Twin-Seat" as the case may be. Connect the oil-pipe union to oil gauge and re-position instrument panel where fitted.

(24) Refill oil-tank, primary chaincase and petrol tank. The motor cycle is now ready for testing.

(25) After testing check all bolts and nuts for tightness.

CHAPTER III

CLUTCH AND GEARBOX

IN order to obtain access to the clutch it is necessary to remove the primary chaincase, outer portion. The foot-rest and the brake pedal must first be taken off. The brake pedal can be taken off the spindle after removing the securing nut. It is not necessary to remove the rod. A large-sized screw-driver is required to remove the chain-cover screws. If necessary the machine should be allowed to lie on the nearside foot-rest, which will permit the mechanic to apply more pressure on the screwdriver. This is important if the slots in the screw heads are not to be damaged. Having removed the screws, the outer portion of the chaincase can be removed. Care should be taken not to damage the paper washer between the inner and outer cases if it is to be used again, a new washer not being available. In no circumstances should a screwdriver or other tool be forced between the two halves of the chaincase in order to separate them, as this procedure is likely to damage the faces. If the outer half is given a sharp tap with a mallet or hammer handle, no difficulty will be experienced in breaking the joint.

It should be noted that there is a tubular distance piece fitted to the foot-rest rod between the case and the frame. If this is not replaced when reassembling, the chaincase may be fractured when the foot-rest rod nut is tightened up.

Dismantling the Clutch

The clutch plates are dismantled by removing the four brass clutch nuts, using the special key provided in the tool-kit. On the underside of the head of each nut is a small " pip " to prevent the nut from unscrewing. When attempting to unscrew the nut, this " pip " may engage with the end of the clutch spring, thus preventing it being turned. In order to overcome this difficulty a knife blade, or a portion of a ground-off hack-saw blade, should be inserted under the head of each nut. This will hold the " pip " away from the spring end while the nut is turned. When the nuts have been taken off their pins, the outer pressure plate and the driven and driving plates can be removed.

Cork Inserts

The corks fitted should protrude $\frac{1}{32}$ in. on either side of the plate. They should be in an oily condition and the steel plates should be smooth and not scored. The tongues on the corked plates should be in good condition and unworn. The free length of the clutch spring should be checked in accordance with the technical data chart. If the length is not the same, or very nearly so, the springs should be replaced.

Clutch Drag

Clutch drag is often caused by sticky plates, and if they are found to be in this condition they should be washed in paraffin or petrol. Sticky plates are often caused by the use of engine oil in the chaincase. The correct grade should be used (SAE 20), but if this is not available a mixture should be made of half engine oil and half paraffin. This mixture is satisfactory for the clutch plates, and is also a good lubricant for the chain. It should only be used if the correct grade is not available. (See lubrication chart, page 211.)

If the clutch cable is not free in the casing, lubricate as described for carburetter control cables, page 154.

Clutch Hub Removal

A withdrawal tool is provided in the tool-kit to facilitate removal of the clutch hub from the gearbox mainshaft after taking off the nut and washers. The hub has an internal thread and the withdrawal tool screws into it. The clutch housing and sprocket revolves on the clutch hub when the clutch is withdrawn. A bearing is provided by twenty ¼ in. by ¼ in. rollers. The bearing surfaces should be examined for wear.

Reassembly of the Clutch

The clutch hub presents no difficulty in re-fitting to the mainshaft. Care should be taken to see that the key is properly positioned and the nut tightened fully up on the washer. The housing and sprocket assembly should be fitted to the hub before it is installed on the mainshaft. Grease should be used on the roller bearing. See that the clutch studs are properly fitted into their slots before assembling the plates.

The number of corked and plain steel plates for each model is shown in the chart below:

Model.	No. Corked Plates.	No. Steel Plates.
2H and 2HC	3	4
3H and 3HW	3	4
5H	4	5
3S and 3SE	3	4
5S and 5SW	4	5
6S	4	5
T70	3	4
T80	3	4
T90	4	5
5T	4	5
6T	5	6
T100 (1951)	5	6
T100	4	5
TR5	4	5

Clutch Plate Lubrication and Fitting

The cork clutch plates should be well soaked in oil or the half-and-half mixture of oil and paraffin before they are assembled. A plain plate is fitted first, followed by a corked plate, then a plain plate and so on, which will leave one plain plate to be fitted last on the outside of the clutch. On the plain plate, the pressure plate is fitted. The clutch rod engages in the button in the centre of the plate. The clutch springs can now be put into their cups and the four nuts screwed into the pins. With new corked plates the nuts should be screwed up until the ends of the pins are level with the heads of the screws. If the corks are worn, a little more pressure on the springs may be necessary.

Final Adjustment of Clutch

The clutch should be extracted by means of the handlebar lever and spun with the kick-starter while the mechanic looks

FIG. 30.—THE CLUTCH ADJUSTMENT.

The clutch rod is adjusted after removing the gearbox filler cap, slacking off the lock-nut and operating the adjuster with the screw-driver. On the top of the gearbox will be seen the cable adjuster which can be turned after slacking off the lock-nut. Also shown is the bottom of the oil tank with the large hexagonal nut which secures the filter to the oil tank. To disconnect the oil feed pipe the smaller hexagonal nut underneath should be removed.

down on to it to make certain that it spins true. If the pressure plate " wobbles " when the clutch is spun an adjustment of the clutch nuts is indicated. By screwing in the nut adjacent to that part of the pressure plate which appears to be nearest the edge of the housing, spinning the clutch and, if still not true, repeating the process, the plate can be made to spin accurately. If the adjustment is not correctly made, the clutch will not free properly, causing a difficult gear change and trouble with selecting bottom gear from neutral.

Cable and Rod Adjustment

After replacing the clutch, the rod and cable adjustment should be checked up. The rod adjustment is exposed by removing the gearbox filler cap. There should be about $\frac{1}{32}$ in. of free movement on the clutch rod. If necessary, adjustment can be made with a screwdriver after slacking off the lock-nut. The cable adjuster is on the top of the gearbox. There should be about $\frac{1}{16}$ in. free movement at the handle-bar lever. When altering the adjustment, do not allow the jaw of the spanner to lever up the adjuster or the lug on the gearbox cover may be fractured.

Gearbox

If the gearbox is to be overhauled it should be removed from the machine. First remove the primary chaincase, clutch and chains as described on page 91. Disconnect the clutch cable at the gearbox lever and remove by unscrewing adjuster from inner cover. Where the speedometer drive is taken from this unit, the drive cable either fits to an external speedometer gearbox, driven from the final-drive sprocket or directly into the gearbox inner cover, where the drive is taken from the layshaft. In both cases the cable is released by unscrewing the nut at the gearbox end and withdrawing the cable from its drive. Now unscrew the nut securing the primary-chain adjuster to the seat tube and then remove the

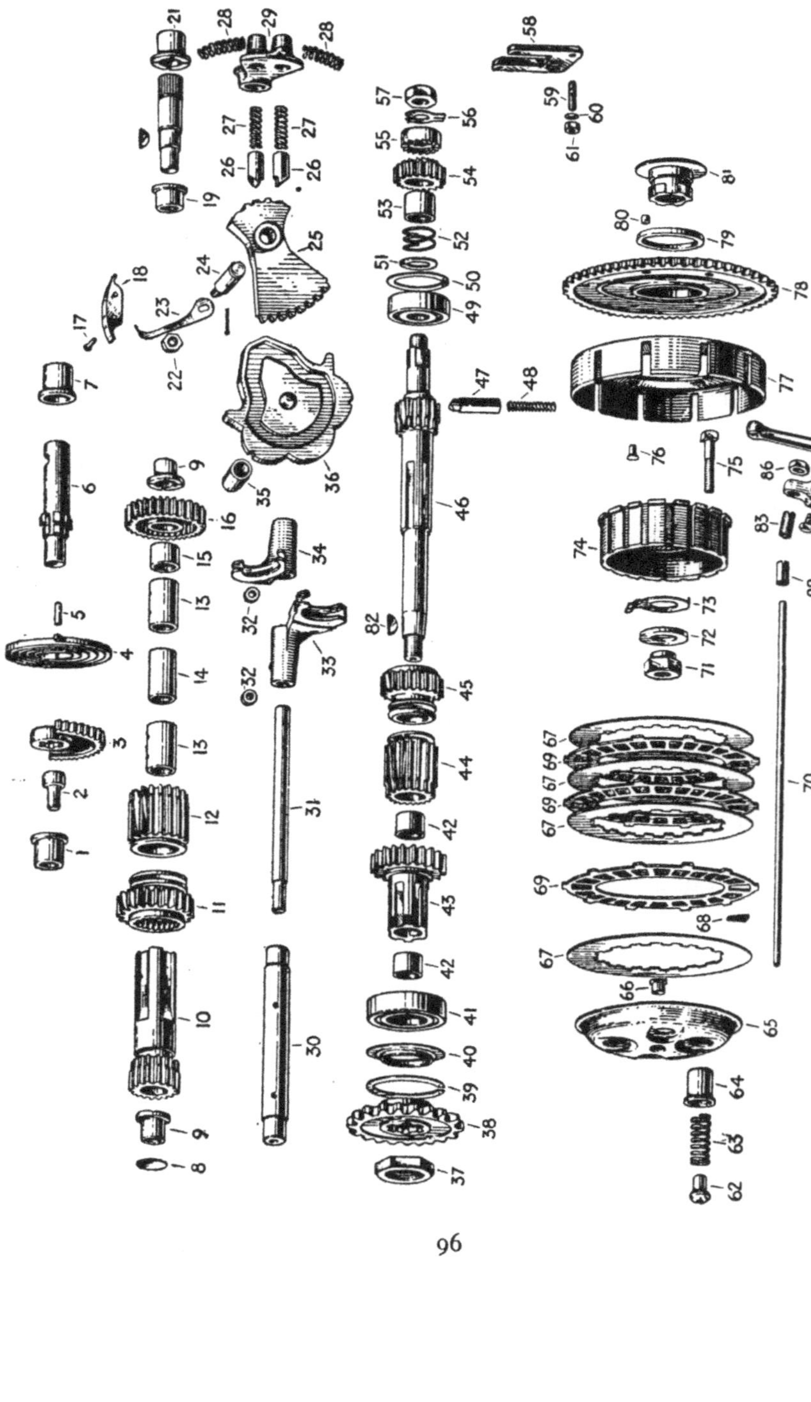

TRIUMPH 4-SPEED GEARBOX (1ST. CONDITION)

FIG. 31.—INTERNAL PARTS OF THE GEARBOX AND CLUTCH. (Key to parts on next page.)

KEY TO PART NOS. IN FIG. 31.

1. T362	12. T436	26. T406	39. E489	50. T280	64. T425	75. T424
2. T45	13. T438	27. T405	40. T445	51. T505	65. T419	76, 77, 78, 79.
3. T25A	14. T432	28. T404	41. T448	52, 53. T488/28	66. T422	T416/428 (500 c.c.)
4. T31	15. T431	29. T408/9	42. T446	54. T730	67. T415	T416A/428 (350 c.c.)
5. T47	16. T437	30. T433	43. T726*	55. T731	68. NT288	80. T394
6. T24A	17. D208-17	31. T291	T465†	56. T732	69. T414	81. T418
7. T23	18. T396	32. NT261	44. T443	57. NT297	70. T455 (350 c.c.)	82. S66-1
8. T18	19. T412	33. T261	45. T727*	58. T407	T455A (500 c.c.)	83. T413
9. T290	21. T57	34. T260	T467†	59. T454	71. T421	84. T401
10. T434*	22. T224	35. T341	46. T441	60. E2312	72. T420	85. T402
T466†	23. T395	36. T500	47. T501	61. F879	73. T524	86. T453
11. T435	24. T399	37. T440	48. T373	62. T427	74. T417 (500 c.c.)	87. T487
T468†	25. T397	38. T471	49. S35-7	63. T429	T417A (350 c.c.)	88. PT108B

* Standard Ratio Gears. † Wide Ratio Gears.

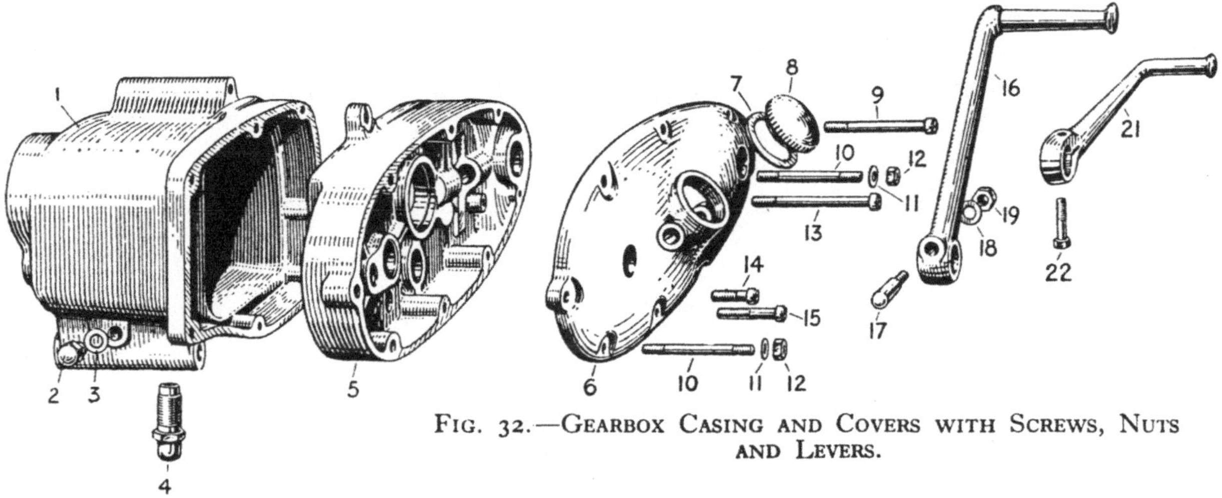

FIG. 32.—GEARBOX CASING AND COVERS WITH SCREWS, NUTS AND LEVERS.

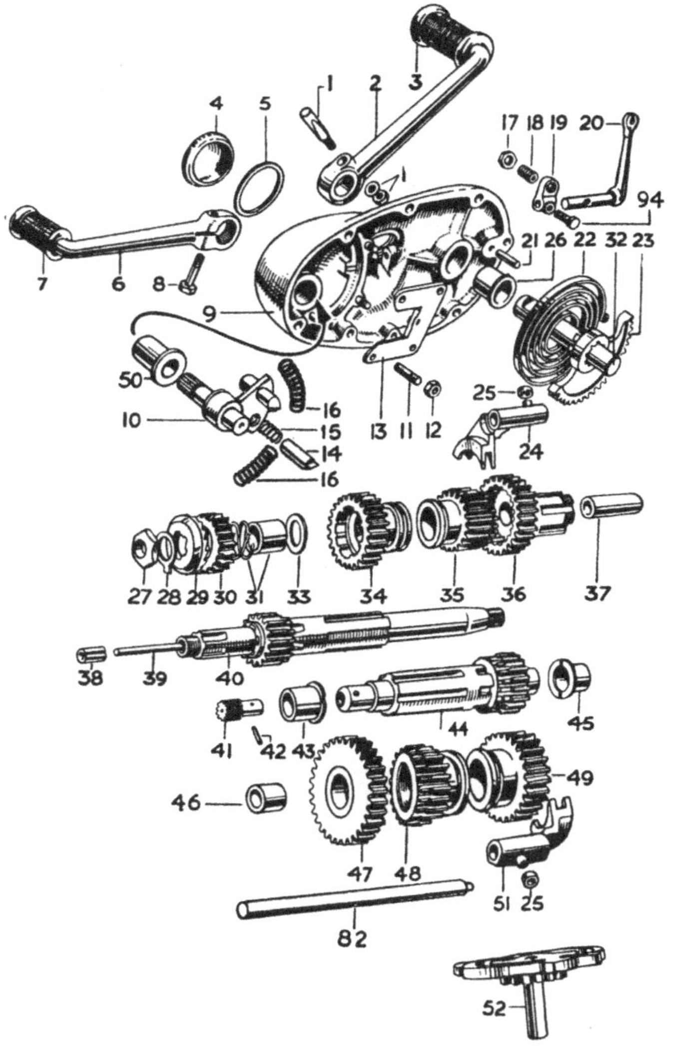

KEY TO PART NUMBERS.

1. S52–1
2. T451
3. F1814
4. T400
5. NT376
6. T398
7. T449
8. T450
9. T935
10. T408/9
11. T454
12. F879
13. T407
14. T406
15. T405
16. T404
17. T453
18. T413
19. T402
20. T487
21. T47
22. T31
23. T25A
24. T260B
25. NT261
26. T23
27. NT297
28. T732
29. T731
30. T730
31. T488/28
32. T24A
33. T505
34. T917
35. T916
36. T918
37. T920
38. PT108B
39. T455
    T455A
    T455B
40. T914
41. T924
42. T925

FIG. 33.—TRIUMPH 4-SPEED

clamping bolt and pivot bolt. The gearbox can now be removed; if there is any difficulty in releasing it from the frame slacken off the adjacent bolt.

KEY TO PART NUMBERS (*contd*).

43. T911
44. T921
45. T911
46. T928
47. T926
48. T930
49. T929
50. T57
51. T913
52. T500
53. T931
54. T45
55. S46–15
56. PT104B
57. T280
58. S35–7
59. T396
60. D208–17
61. T224
62. S25–3
63. T395
64. T933
65. T948
66. T934
67. T825
68. T412
69. T323A
70. E698
71. S46–15
80. PT104B
81. T909
82. T291
83. T440
84. T501
85. T373
86. T372
87. T341
88. T448
89. T946
90. T489
91. T471A
92 T397
93. T399

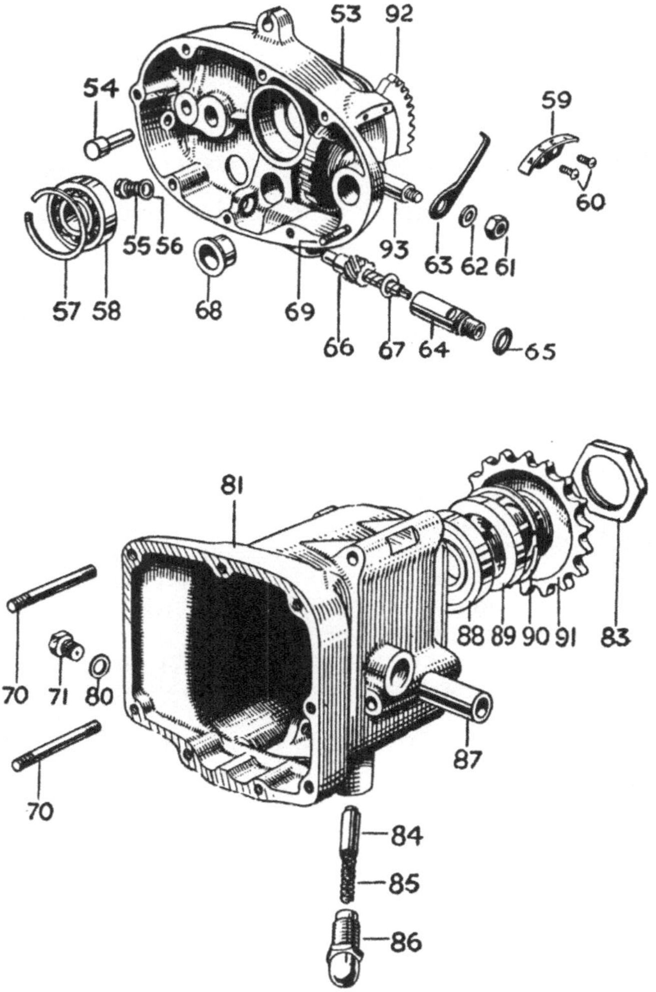

GEARBOX (2ND CONDITION).

As a more robust gearbox had to be designed for the "Thunderbird" (6T), this was standardised throughout the range of models for 1950. The foot-change and kick-starter mechanism remains the same, but all gears are completely new; therefore, to assist the reader, each pattern will be dealt with separately with the exception of the afore-mentioned items.

Dismantling Gearbox (1st Pattern 1937/49)

Clamp the gearbox in a vice by the ends of the bottom lug with the gear change end facing the mechanic. The sequence of dismantling after draining the oil is as follows:—

(1) Remove the nut on the mainshaft (drive side). In order to prevent the shaft turning use a small piece of metal bar about 2½–3 in. long, placing one end on the jaw of the vice and the other under a tooth. Do not remove the sprocket at this stage.

(2) Remove five screws and three nuts securing the outer cover to the gearbox casing.

(3) Take off the cover by gripping the foot-change lever and the kick-starter crank and rotating the crank a little to the left (mechanic facing the box). The joint is made with jointing compound so that the faces will be stuck together, thus requiring a little force to break them.

(4) Remove the kick-starter ratchet nut, preventing the shaft from turning with the metal bar as described in paragraph (1) above.

(5) Slide the ratchet mechanism from the shaft. The teeth of the ratchet and pinion should be unmarked and clean cut. If they are badly worn, the kick-starter will not engage or will slip badly. The spring should not be distorted and the sleeve unworn.

(6) Remove the three screws securing the inner cover to the gearbox shell.

(7) Take off the inner cover by tapping the drive end of the mainshaft with a lead hammer or a piece of wood. The main-shaft will then come away with the cover.

(8) Tap out the selector fork spindle with a long ¼-in. drift.

(9) Remove the gearbox from the vice, grip a ⅜-in. bar in the jaws of the vice and turn the gearbox with the open end uppermost, fitting the ⅜-in. bar into the swivel boss.

(10) Remove the gears, selectors, cam-plate and plunger from the casing. There are two small steel rings which fit over the pegs on the selector forks. Care should be taken that these are

not lost. It should be noted that there is a drain plug in the rear of the case. Draining should not be carried out by removing the acorn nut which secures the cam-plate plunger and spring. Where the plunger and spring are found to be missing, this is often due to careless draining.

(11) The ball bearings are held in position by means of circlips, which can be removed with sharp-nosed pliers. After removing the circlip, the bearing can be tapped out.

(12) The layshaft bushes can be driven out with a soft drift

FIG. 34.—THE KICK-STARTER RATCHET ASSEMBLY.

of suitable size. The bush in the case has a dust-cover disc which should first be tapped out from inside the box. New bushes need not be reamered after fitting.

(13) The layshaft and mainshaft high gear bushes should be examined for wear. When new bushes are fitted no reamering is necessary, as an allowance is made for pressing in during manufacture. It should be noted that a distance piece is fitted between the two bushes in each gear.

(14) Examine all gears and other parts and replace any not serviceable. The gear-change lever return springs should be particularly inspected, and it should be seen that they are approximately each the same length. The foot-change plunger

springs must be replaced if doubtful as regards strength or if they are rusty.

Kick-Starter Crank Return Spring

If the return spring is to be replaced the gearbox need not be removed from the machine.

(1) Remove the outer gearbox cover as described in (2) and (3), page 100.
(2) Take off the crank and then the spindle and quadrant. Note the spline to which the inner end of the spring is fitted.
(3) Remove the old spring and fit a new one to the same spline, attaching the outer end to the peg in the cover. Note the position of the spring in the illustration. When the crank is depressed the spring should wind up—not unwind. Too much tension in the spring will cause it to break. Too little tension will cause the pinion to engage. Errors in tension are caused by fitting the inside end of the spring to the wrong spline on the shaft.
(4) Fit the crank.
(5) Fit the cover as described in (14), page 105.

Foot-Change Gear Control Mechanism

Faulty gear changing is sometimes caused by the quadrant plungers sticking due to weak springs or the plungers being gummed up. The plungers should be polished and move freely in the quadrant housing. If the spring is suspected the free length can be checked up from the data table (page 206). The foot-change return springs should also be checked, as any weakness or failure here will prevent the pedal from returning to its neutral position and cause incorrect positioning of the plungers.

To dismantle the foot-change gear control mechanism, proceed as follows:—

(1) Remove outer cover as described on page 100.
(2) Take off foot-change pedal by unscrewing the set screw and easing the pedal off the splined shaft.
(3) Remove the guide plate by unscrewing the four nuts and withdraw the control mechanism.

Assemble the foot-change gear control mechanism as follows :—

(1) Fit the quadrant to the cover.

(2) Allow the quadrant to drop to its lowest position and install one of the radiused springs. Now apply pressure to the quadrant so that the spring fitted is compressed and position the other.

(3) Fit the plungers and springs to the quadrant. The plungers should be assembled with the ground-angled faces pointing towards each other.

(4) Place the guide plate over the four studs with the bevelled edge downwards. Depress the plate so that the cutaway part of the plunger protrudes and screw up the four nuts.

(5) Replace foot-change pedal.

Reassembling the Gearbox

Immediately prior to reassembly, make certain that all parts are scrupulously clean and before each part is placed in position lubricate it well with engine oil. Proceed in the following sequence :—

(1) With the gearbox casing on the bench fit the cam-plate into the case with the pointed end towards the closed side of the casing (the clutch end). This will place the cam-plate in the high gear position. Insert and screw up the cam-plate plunger assembly.

(2) Insert the mainshaft high gear through the bearing in the case and screw on the nut a few threads to keep it in position.

(3) Place the gearbox in the vice, fitting the swivel lug on to the bar as described in (9) (page 100).

(4) Insert the remainder of the mainshaft gears, the mainshaft selector and finally the mainshaft itself. (See Fig. 35 for correct position.)

(5) Insert the layshaft gears (less the low gear and the layshaft) and layshaft selector. (See illustration for correct position.)

Note.—The mainshaft selector is the smaller of the two. Do not omit the two steel rings on the selector pegs. Use grease to keep them in position while assembling.

(6) Fit the selector shaft.

(7) Insert the layshaft.

(8) Put the layshaft low gear on the layshaft.

(9) Replace the inner cover, using jointing compound to make the joint.

FIG. 35.—LAYSHAFT AND MAINSHAFT GEARS.
(1ST CONDITION).

FIG. 36.—INTERIOR OF THE GEARBOX OUTER COVER.
(1ST CONDITION).

Showing the kick-starter quadrant and return spring and the selector plungers showing through the plate. The gear change return springs are also visible at the top and bottom of the plate.

(10) Tighten evenly the screws fixing the cover to the case.

Note.—When the inner cover is fitted the gearbox must be in top gear. If the cam-plate has been correctly inserted this should present no difficulty. Check up on the gear indicator on the top of the gearbox before tightening the cover screws.

(11) Remove the box from the bar and clamp in the vice as in dismantling (page 100).

(12) Replace the kick-starter parts on the shaft and tighten up the nut. Use a new tab washer to retain it in position. (See Fig. 36.) Test the ratchet for freedom.

(13) Fit the final drive sprocket and tighten the nut; use a new tab washer.

Note.—To facilitate tightening the nuts, use the small bar described on page 95. Where tab washers are not available, the nuts should be centre-punched in three places.

(14) Assemble the outer cover to the box. Use jointing compound to make an oil-tight joint, and position the kick-start crank half-way down before pushing the cover into position.

(15) Tighten up the screws and nuts, progressing evenly.

(16) Test the gear selection.

(17) Adjust the chains (see page 110).

(18) Assemble gearbox to machine.

(19) Re-fit the clutch and fit chains.

(20) Adjust the clutch (see page 94).

(21) Re-fit primary chaincase, both sections, foot-brake, foot-rests, etc.

(22) Fill the primary chaincase and gearbox with the correct amount and grade of lubricant (see page 211).

Dismantling Gearbox (2nd Pattern)

(1) Clamp the gearbox in the vice by the ends of the bottom lug, gear change facing the operator.

(2) Select bottom gear.

(3) Remove outer cover, three screws and three nuts. Take off this cover by gripping the foot-change lever and the kick-starter crank; the crank should be rotated a little to the left. The joint is made with jointing compound, therefore a little force will be required to break it.

(4) Remove the nut securing sprocket to high gear. To do this use a small piece of tube or bar about 2½ to 3 in. long, placing one end on the jaw of the vice and the other under the sprocket tooth in order to prevent the sprocket turning when force is applied to the spanner. Tap the sprocket off the high gear.

(5) Slide the high gear along the mainshaft towards the

front of the box ; this will lock all gears to enable the kick-starter ratchet-nut removal.

(6) Remove kick-starter ratchet nut and withdraw kick-starter ratchet assembly.

(7) Take off the gearbox inner cover—four screws. Note, one screw is situated inside the cover adjacent to the selector quadrant.

(8) Withdraw the selector rod and mainshaft.

(9) The gear cluster with selectors can now be taken out of the casing ; take care not to loose the cam-plate rollers fitted to the selectors.

(10) The mainshaft high gear which remains in the bearing can be taken out.

(11) Withdraw the cam-plate and remove the cam-plate plunger and spring.

(12) Both bearings can be removed by taking out the circlips with sharp-nosed pliers and tapping them away from their housings.

(13) If it is necessary to remove the speedometer-driven gear from the outer cover, this operation can be carried out by drifting the gear complete with bush from the inside of the casing. Use a soft-metal drift to avoid damaging the gear.

Reassembling the Gearbox

(1) Refit the bearings to their respective housings and lubricate.

(2) Replace the cam-plate and insert and screw up the cam-plate plunger assembly.

(3) Insert the mainshaft high gear and tap on sprocket, do not fit the nut.

(4) Place a $\frac{3}{8}$-in. bar upright in the vice and turn the gear-box with the open end uppermost, fitting the bar into the swivel boss.

(5) Place the mainshaft second and third gear together, selector groove facing inwards and then fit selector with cam-plate roller over the second gear (long). In order to hold the roller in position it should be well greased.

(6) Insert the assembly into the casing, second gear meshing with the high gear and the cam-plate roller engaging in the cam-plate track.

(7) Assemble the layshaft second and third gears to the layshaft high-gear splines, large gear first, selector grooves towards the centre. To this assembly fit the selector fork with the roller facing the speedometer-driving gear.

(8) Fit the layshaft cluster in the casing, again engaging the cam-plate roller with the cam-plate track. Enter the selector spindle through the selector forks, engaging the small diameter in the locating hole in the clutch end of the casing.

(9) Fit the mainshaft and low gear through the mainshaft cluster and layshaft low gear over the spindle of the layshaft engaging with layshaft third gear.

(10) Remove assembly from the vice and replace again as when dismantling, open end facing the operator.

(11) Smear a film of jointing cement on tne inner face of the inner cover and assemble cover to casing with the indicator in the low-gear position.

(12) Fit the screw adjacent to the selector quadrant and test the gear selection. This can be done by operating the quadrant with a screwdriver; turn high gear and mainshaft when testing.

(13) Replace the remaining screws to the inner cover and then fit the kick-starter assembly—plain washer, spring, pinion, ratchet lock washer and nut. Lock gears as previously stated when tightening the nut.

(14) If the speedometer-driven gear has to be replaced, a special tool is necessary to guide the rubber oil-seal into position. The gear should first be threaded through the aperture and then the bush with oil-seal driven into the casing. Ensure that the slot in the bush is aligned with the screw hole in the casing.

(15) Assemble the outer cover complete with foot-change mechanism to the gearbox. Use jointing cement on the faces and position the kick-start crank half-way down before pushing the cover into position. Tighten the cover screws and nuts.

(16) Fit the nut to the chain sprocket, locking the sprocket with a bar, as when dismantling, in order that the nut can be tightened. Centre punch the nut on the thread end to lock.

CHAPTER IV

BRAKES, CHAINS, WHEELS AND HUBS

BOTH brakes are adjusted by means of knurled hand adjusters. The pedal position of the rear brake can be adjusted on the stop fitted to the frame.

Brake Wear

The brake linings on the rear-brake shoes normally require renewing if the brake cam lever is approaching the vertical with brake off and the rear wheel on the ground. As the adjustment is taken up on the front brake, more and more of the adjuster thread will be exposed. The linings should be renewed when about ½-in. of thread can be seen. The fitting of new linings presents no difficulty. Access to the brake parts is obtained after removing the wheel by taking off the spindle nut and brake anchor plate locking nut, when the anchor plate can be removed with the brake parts attached. The linings

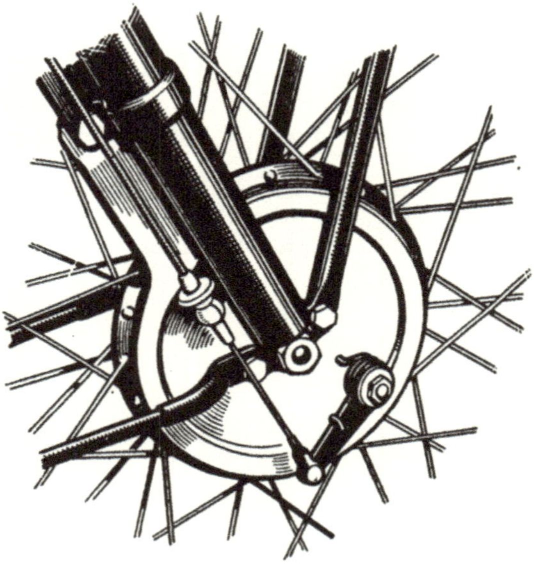

FIG. 37.—BRAKE ADJUSTMENT ON MACHINES FITTED WITH TELESCOPIC FORKS.

are supplied ready drilled and countersunk for the rivets and are the correct shape to fit the shoes. After fitting, the sharp edges should be eased off and the ends chamfered with a file.

Greasy Linings

If oil or grease has found its way on to the linings, it must be removed. No amount of washing in petrol will eliminate grease if the lining is impregnated. Though it may appear to have been removed, more grease will exude when the brake heats up in operation.

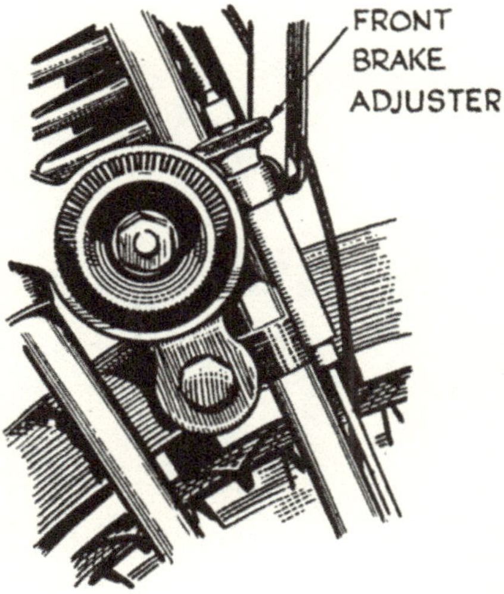

Fig. 38.—Front Brake Adjuster on Machines Fitted with Girder Forks.

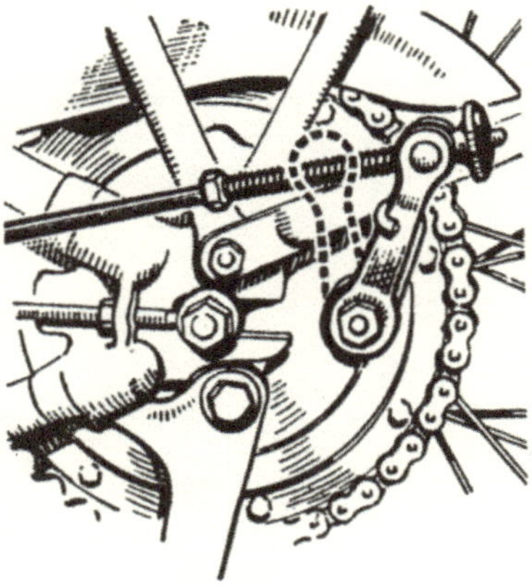

Fig. 39.—Position of the Rear Brake Lever with the Brake Linings in Good Condition.

With worn brake linings the lever will take up the position as shown by the dotted line.

CHAINS

Care of Chains

The rear chain should be removed for thorough cleaning and lubrication every 2,000 miles. The primary chain does not require attention so frequently, as it is totally enclosed.

If the primary chain is removed for any reason it should be cleaned. Brush with a stiff brush in a paraffin bath and wash in clean paraffin. Hang up to drain. Melt some grease in a suitable container and put the chain into it. The grease should be just melted: care must be taken not to overheat it. See that the chain is entirely covered and leave it in the grease until this has solidified again. Now remove the chain and wipe off the superfluous grease. If the chain is

FIG. 40.—TESTING THE ADJUSTMENT OF THE REAR CHAIN.
The gear should be in neutral. This illustration also shows the rear wheel adjuster and lock-nut.

removed before the grease has solidified the melted lubricant will drain out. See page 91 for method of removing primary chaincase.

Tension—Rear Chain

There should be approximately ½ in. up-and-down "slack" in the middle of the run of the rear chain. To adjust, slacken off the two spindle nuts and also the chain adjuster lock-nuts. Turn the chain adjusters an equal number of turns on each side in order to keep the wheel in true line. Test the chain

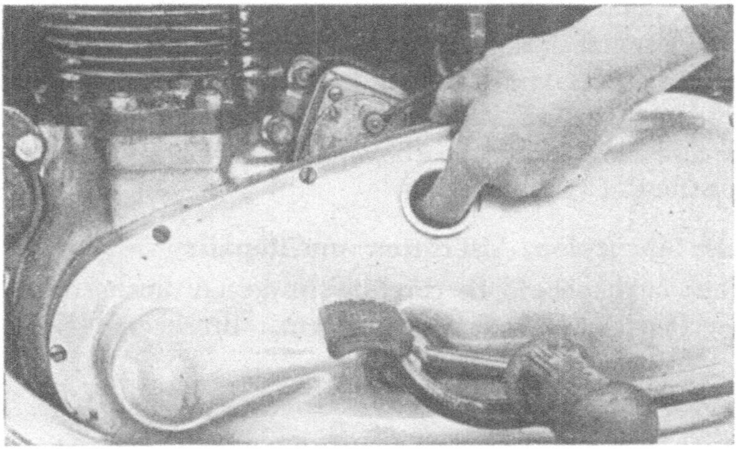

FIG. 41.—TESTING THE ADJUSTMENT OF THE PRIMARY CHAIN AFTER REMOVING THE CHAINCASE FILLER CAP.

in a number of different positions, as the amount of slack may vary owing to uneven stretch. Adjustment for tension may make necessary the adjustment of the rear brake.

Tension—Primary Chain

The filler cap on the primary chaincase should be removed to inspect the chain. There should be ½ in. total up-and-down " slack ". To adjust, the gearbox must be moved on the

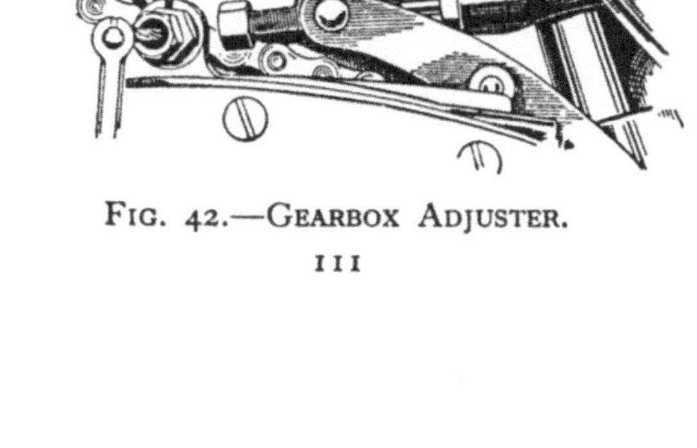

FIG. 42.—GEARBOX ADJUSTER.

swivel. Slacken off the gearbox clamping bolt and the swivel bolt, loosen the lock-nut on the gearbox adjusting bolt, and make the necessary adjustment by turning the bolt. Finally, tighten up the lock-nut and the clamping and swivel bolts. The alteration in the gearbox position will necessitate the adjustment of the rear chain.

Chain Inspection, Alteration and Repairs

The chains should be carefully inspected during overhaul to see that all the rollers are present. Breakage of primary

FIG. 43.—PORTIONS OF THE CHAIN REFERRED TO IN THE TEXT.

chain rollers will cause very rough running and also rapid wear on the sprocket teeth. Maximum permissible stretch is 3 per cent. of the length.

If the chains have been correctly serviced very few repairs should be necessary. But should the occasion arise to repair, lengthen or shorten a chain, a rivet extractor and a few spare parts will cover all requirements.

To *shorten* a chain containing *an even number of pitches* remove the dark parts shown in " A ", and replace by cranked double link and single connecting link as " B ".

To *shorten* a chain containing an *odd number of pitches* remove the dark parts shown in " C " and replace by single connecting link and inner link as " D ".

To *repair* a chain with a broken roller or inside link remove the dark parts in " E " and replace by two single connecting links and one inner link as " F ".

Rivet Extractor

The rivet extractor can be used on all motor cycle chains up to ¾ in. pitch, whether the chains are on or off the wheels.

When using the extractor :—

(1) Turn screw anti-clockwise to permit the punch end to clear the chain rivet.

(2) Open the jaws by gripping tommy bar and handle together. (See illustration.)

(3) Pass jaws over chain and release grip. Jaws should rest on a chain roller free of chain-link plates. (See illustration.)

(4) Turn screw clockwise until punch contacts with and pushes out rivet end through chain outer link plate. Unscrew punch, withdraw extractor and repeat complete operation on the adjacent rivet in the same chain outer link plate. The outer plate is then free and the two rivets can be withdrawn from the opposite side with the opposite plate in position. Do not use the removed part again.

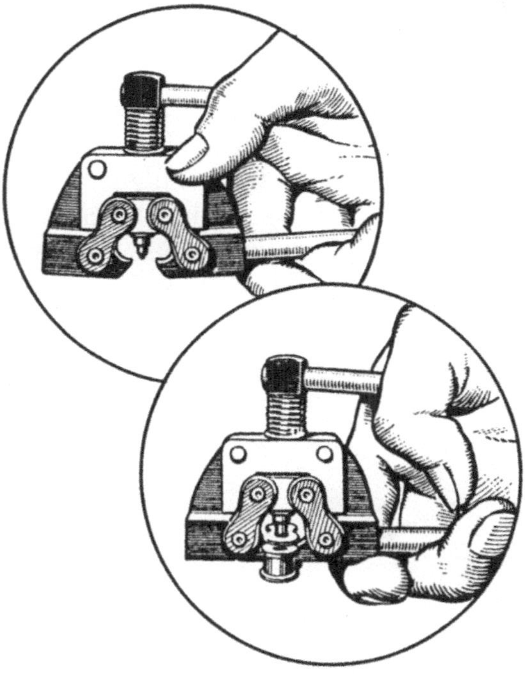

FIG. 44.—CHAIN RIVET EXTRACTOR.

WHEELS AND HUBS (PRE-WAR)

Front Wheel Removal and Replacement

Disconnect the speedometer drive and front-brake rod at the lower end. Remove the brake torque-rod bolt. Lower the front stand and slacken off the spindle nuts, when the wheel can be withdrawn. To replace, reverse the removal procedure.

Rear Wheel Removal

Slacken off the nuts securing the tailguard mudguard stays and remove the two hexagonal-headed screws securing the tailguard to the mudguard blade. Where pannier equipment is fitted the steady stay on the tailguard should be detached from the pannier frame at each end. Remove the tailguard and place it at the side of the machine. It is not necessary to remove the rear-lamp cable as there is sufficient slack to allow for the removal of the tailguard. Break the rear chain by removing the spring link. Be careful to see that the gearbox is not in the neutral position, as if the gearbox sprocket is free the weight of the chain may rotate it and the chain will then fall off. Screw off the brake-adjusting nut. Slacken off the two spindle nuts and withdraw the wheel from the fork.

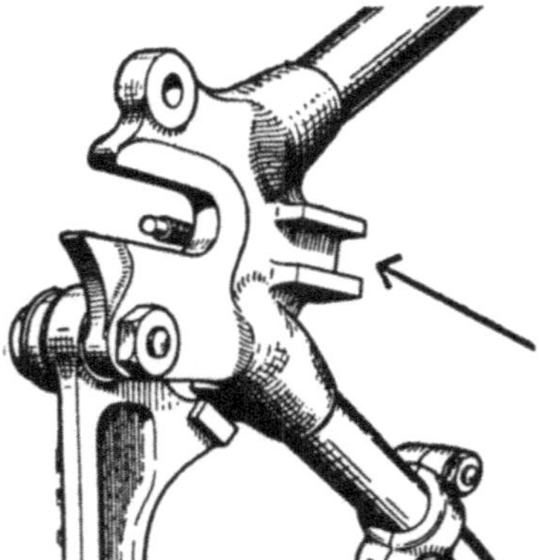

FIG. 45.—THE CHANNEL IN THE REAR PORTION OF THE FRAME WHICH TAKES THE STUD ON THE REAR ANCHOR PLATE SHOWN IN FIG. 46.

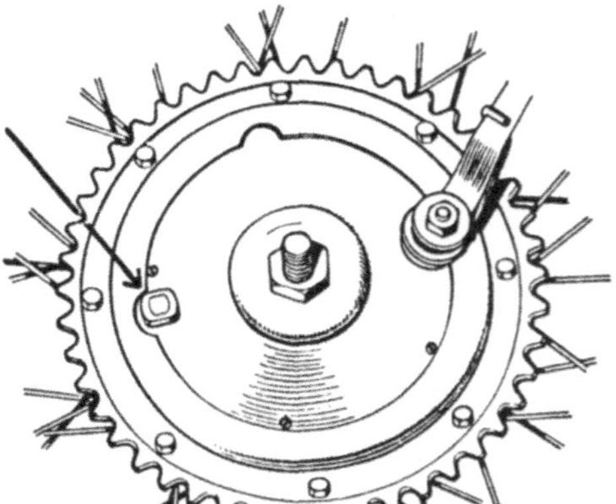

FIG. 46.—THE STUD ON THE REAR BRAKE ANCHOR PLATE WHICH FITS INTO THE CHANNEL SHOWN IN FIG. 45.

Rear Wheel Replacement

To replace the wheel, reverse the removal procedure. Note that there is a stud on the brake anchor plate which must be fitted into the locating channel on the inside of the rear fork.

Bearings—Dismantling and Assembly

The order of dismantling and assembly will be clear from Fig. 47. After removing the parts from one side of the spindle it can be tapped out. When assembling, the hub should not be packed full of grease. The bearings themselves should be copiously packed with grease, when further attention will not be required for at least 10,000 miles. (See page 211 for details of wheel-bearing lubricant.)

Adjustment of Wheel Bearings

Great care must be taken in adjusting the taper roller wheel bearings after assembly of the wheel to the machine. If the adjustment is too close the bearings will quickly give trouble. On the other hand, a properly adjusted and lubricated bearing will give very long service without any further attention.

There is a bearing-adjusting nut secured by a locking nut on each spindle at the side opposite to the brake mechanism. The lock-nut should be slackened off and the adjusting nut turned until it is right home. Then slacken off a little until there is just the slightest amount of lateral movement perceptible when the rim is gripped with both hands at opposite points and rocked by pushing with one hand and pulling with the other. This operation must be carried out with the machine on the front or rear stand. Finally, spin the wheel to make certain that it is quite free. If it does not spin easily, remember the brake may be the cause of the trouble.

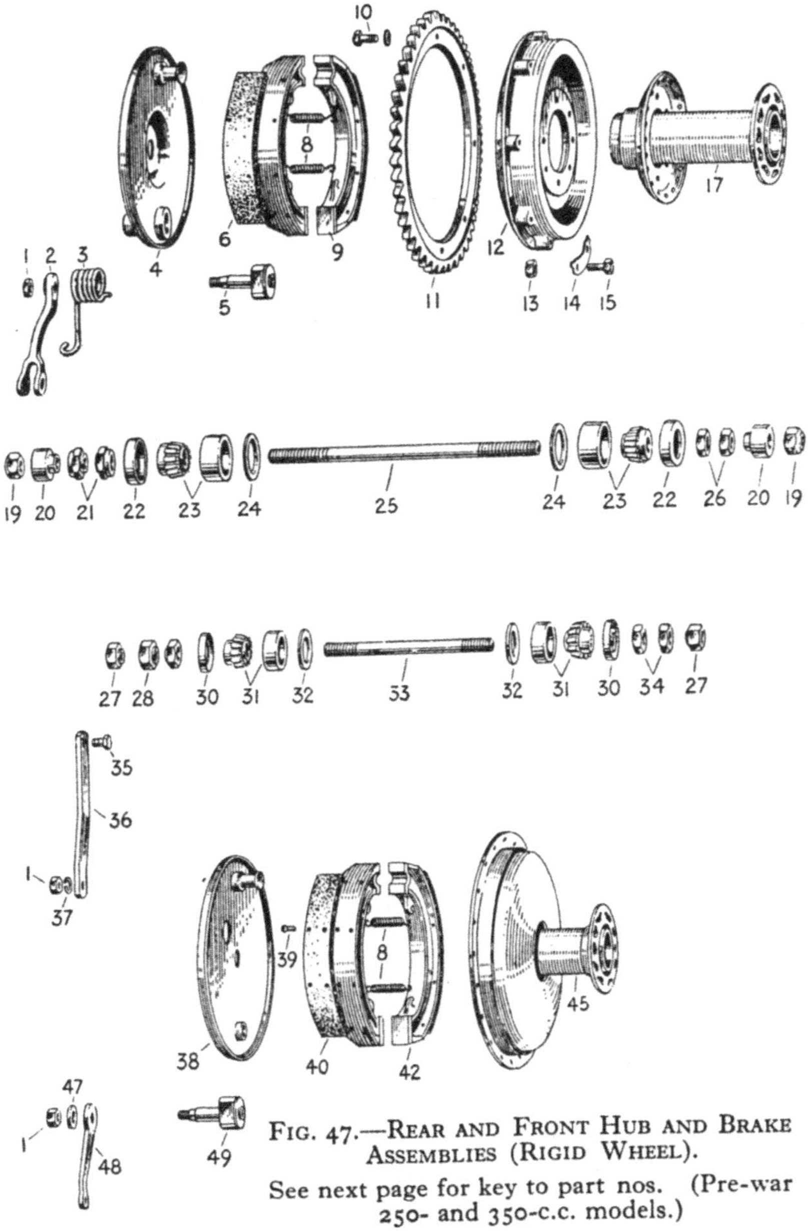

FIG. 47.—REAR AND FRONT HUB AND BRAKE ASSEMBLIES (RIGID WHEEL).

See next page for key to part nos. (Pre-war 250- and 350-c.c. models.)

KEY TO PART NOS. IN FIG. 47.

1. S1–52	12. W311	22. W12	33. W323	42. W343
2. W74A	13. S1–51	23. W7A	34. W192	43. NW256
3. W123	14. W63	24. NW225	35. NW258	44. W338
4. W101	15. W62	25. W61	36. W345	45. W321
5. NW234	16. W57	26. W4	37. SCX38	46. W212
6. NW243	17. W58A	27. W175	38. W314	47. S25–6
8. W135	18. NW222	28. W346	39. W129	48. W313
9. W93	19. TS42	30. W202	40. W344	49. W386
10. W64	20. W6	31. WW11	41. W342	50. W67
11. W56	21. W5	32. W190		

WHEELS AND HUBS (POST-WAR)

Front Wheel Removal

(1) Place the machine on the rear stand.

(2) Remove the pivot pin from the lower brake lever and disconnect brake adjuster.

(3) Remove the brake torque stay bolt on the off-side member and take off the wheel spindle nut.

(4) Raise the machine on to the front stand.

(5) Slacken off the pinch bolt in the near-side spindle lug and tap out spindle from the off-side ; withdraw wheel.

Front Wheel Replacement

(1) Position the front wheel between the fork members and enter wheel spindle from the near side.

(2) Tap the spindle through until the threaded portion is visible through the opposite fork lug. Use a soft metal tool when carrying out this operation, otherwise the boss on the spindle will be distorted, causing the spindle not to enter the lug correctly.

(3) Refit the torque stay bolt and tighten up.

FIG. 48.—TESTING THE REAR WHEEL BEARING ADJUSTMENT.

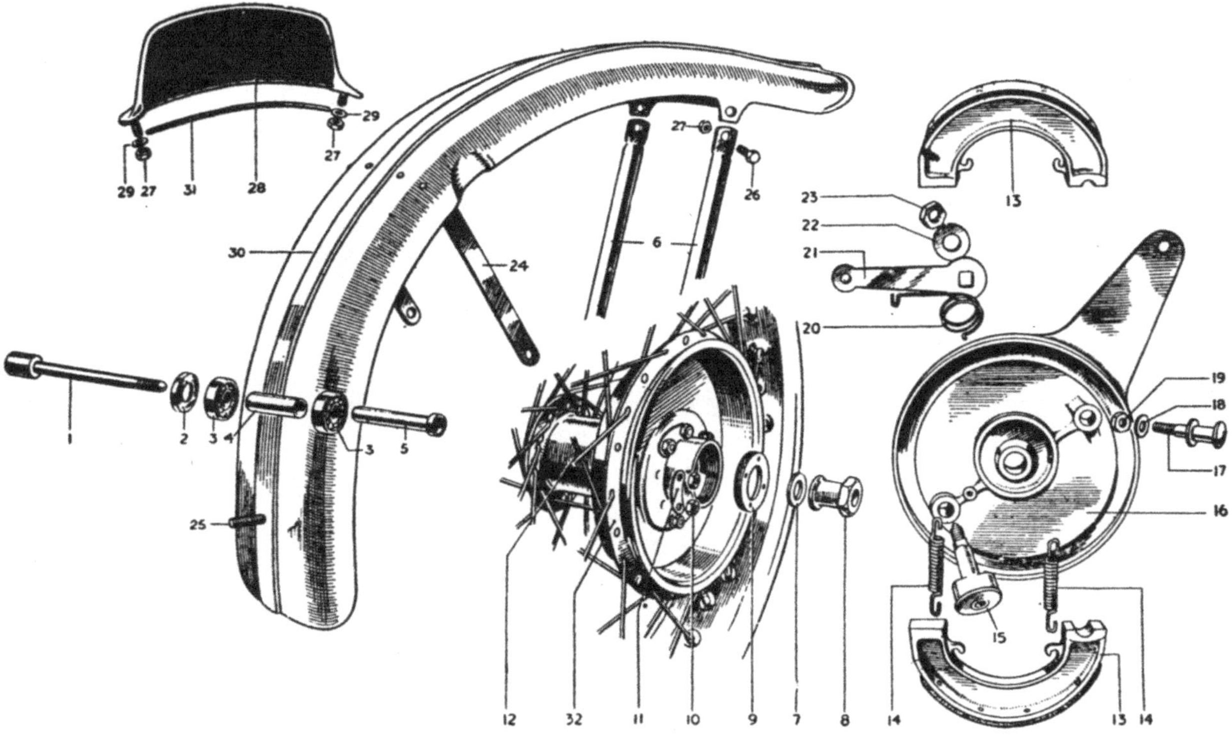

FIG. 49.—FRONT WHEEL AND BRAKE ASSEMBLY WHERE TELESCOPIC FORKS ARE FITTED.
Key to parts is given on next page.

(4) Fit the spindle nut, but do not fully tighten.
(5) Re-position front stand.
(6) Now grip the handlebars and depress the fork movement three or four times. This procedure ensures the correct alignment of the lower fork members to the fork stanchions and centralises the front wheel. *It is essential to fit the front wheel by the above method, otherwise the wheel will be out of alignment and the steering adversely affected.*
(7) Tighten up the spindle pinch bolt and spindle nut and refit front brake attachments.

Rear Wheel Removal (Rigid)

(1) Place machine on rear stand.
(2) Slacken off the nuts securing the tailguard stays to the frame and remove the two hexagonal-head screws fixing the tailguard to main guard. The rear-lamp cable need not be disconnected, as there is sufficient slack to allow for removal of the tailguard. (On 1949/51 5T and T.100, the rear mudguard is detached by removing two nuts under saddle, unscrewing stay to frame bolts and stay to chainguard bolt. Tail-lamp wire is disconnected at mudguard joint.)
(3) Remove the spring link from the rear chain. Before breaking the chain ensure that the gearbox is not in neutral position, for if the gearbox sprocket is free the weight of chain may rotate it, causing the chain to fall off.
(4) Screw off the brake-adjusting nut.
(5) Disconnect the speedometer drive cable at the rear wheel speedometer gearbox where fitted.
(6) Slacken wheel nuts and withdraw wheel from rear stays.

Rear Wheel Replacement

To replace the wheel, assemble in the reverse procedure to removal. Careful note should be made when engaging the stud on the brake anchor plate to the locating channel on the inside of the near-side rear stay (see Figs. 45 and 46).

KEY TO PART NOS. IN FIG. 49.

1. W578	8. W544	14. W135	20. W592	26. DS57
2. W583	9. W582	15. W386A	21. W589	27. F879
3. W653	10. S1-51	16. W585	22. S25-6	28. F2698
4. W581	11. W63	17. W588	23. S1-52	29. S26-2
5. W761	12. W573A	18. S25-2	24. H427	30. H566
6. H569	13. W93	19. PO172A	25. DS55	31. H251
7. W545				

Wheel Bearings—Dismantling and Assembly

The sequence of dismantling and assembly of the front and rear wheel bearings is illustrated on pages 116 and 121. These illustrations will clearly aid the fitter to assemble the wheels correctly.

Greasing Hub Bearings

It will be noted that on Triumph machines manufactured on and after 1945 and most Army models, grease nipples have been omitted from the wheel hubs. This has been done to prevent owners over-greasing the hubs, causing grease to pass the seals and enter the brake drum, which may have disastrous results. When assembling the hub the bearings only should be copiously packed with grease. The wheel should then be periodically dismantled and bearings repacked every 20,000 miles.

Bearing Adjustment

The front-wheel bearings are of the ball-bearing type and will, therefore, require no adjustment. Method of detecting wear is shown in the illustration on page 117. The rear wheel bearings, however, are of the taper roller type, and care must be taken when adjusting after assembly, or at any other subsequent time. If the adjustment is too close the bearings will quickly give trouble. On the other hand, a properly adjusted and lubricated bearing will give very long service without any further attention. Where the rear speedometer drive is fitted it will be necessary to remove the rear wheel and disconnect the speedometer gearbox in order to gain access to the bearing adjustment nuts. The adjusting nut should be turned until it is right up against the bearing and then slackened off a little until there is just the slightest amount of lateral movement perceptible. After the wheel has been re-installed

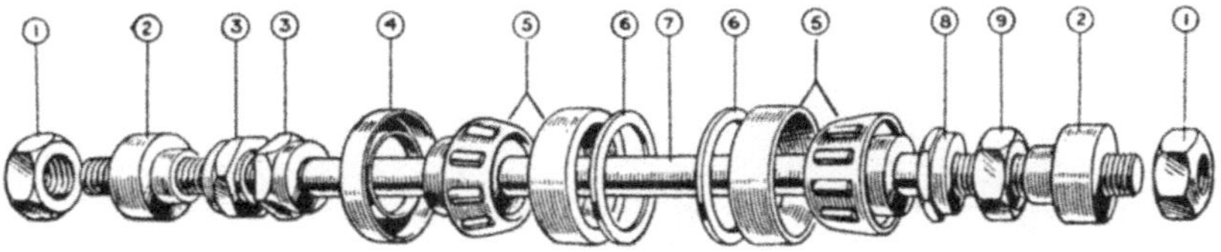

FIG. 50.—THE REAR WHEEL AXLE ASSEMBLY (RIGID WHEEL).

Part Nos. 1. TS42; 2. W6; 3. W5; 4. W12; 5. W7A; 6. NW225; 7. W61; 8. W597; 9. W4.

This shows axle assembly when speedometer gearbox drive is fitted between Nos. 8 and 9. When drive is taken off gearbox chain sprocket, No. 8 is deleted and another No. 9 is added. Dust cover No. 4 should also be added between Nos. 5 and 8.

in the frame a final test of the lateral movement should be made. This can be done by gripping the wheel with both hands as shown in Fig. 48, and if the wheel is alternately pulled and pushed the amount of lateral movement from the axis is discernible. Finally, spin wheel to make certain that it is quite free. If it does not spin easily, check the brake adjustment, as the brake lining may be binding.

SPRING WHEEL

Maintenance

Wear on the bearing is negligible, but adjustment may become necessary (Mark I only). If some lateral play is detected at all positions round the rim, the bearing should be adjusted as described later in this section.

There must be working clearance between the sliding portions of hub; this can be felt at the rim in two opposite positions only. This is quite in order and must not be confused with mal-adjustment of the bearings.

In no circumstances should the plunger box be dismantled without use of a special jig as described on page 125, as otherwise the fitter may be very badly injured when box is parted.

KEY TO PART NOS.
FIGS. 51 AND 51A.

1. W849
   W899 (Mk. II)
2. W601
   W909 (Mk. II)
3. W744
4. W611
   W901 (Mk. II)
5. W609
   W897 (Mk. II)
6. W610
   W904 (Mk. II)
7. W886
8. W885
9. W638
10. W602A/3A
    W906/7 (Mk. II)
11. W664A
12. W658A
13. W600
    W905 (Mk. II)
14. F879
15. E2312
16. W890
17. W766
18. W616
    W908 (Mk. II)
19. S1-40
20. GS299 (Mk. II)
21. T450
    E1562 (Mk. II)
22. W859
    W903 (Mk. II)
23. NW100
    H687 (Mk. II)
24. W635
25. W633
26. W798
27. W797
    W900 (Mk. II)
28. W799
30. W794
    W898 (Mk. II)
31. W791A
    W895 (Mk. II)
32. W836
33. W837
34. W634

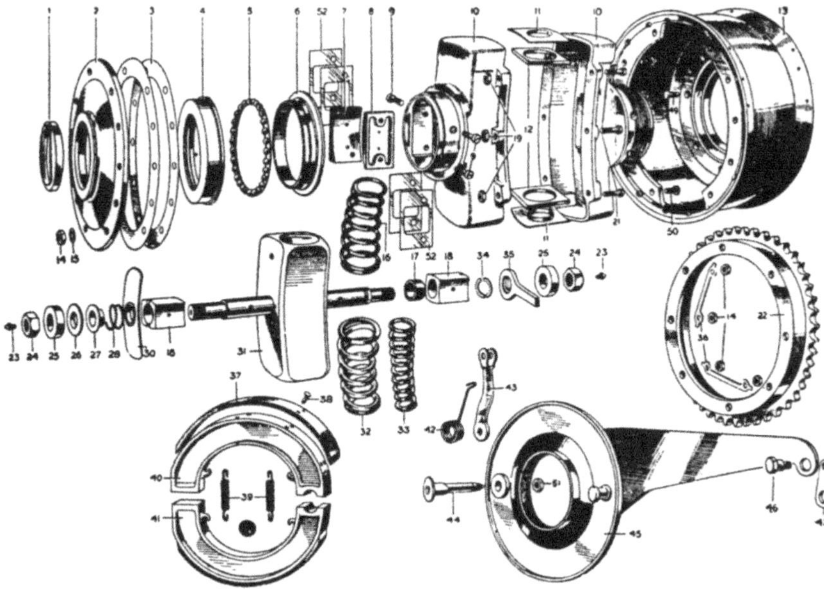

FIG. 51.—TRIUMPH SPRING WHEEL, MARK I.

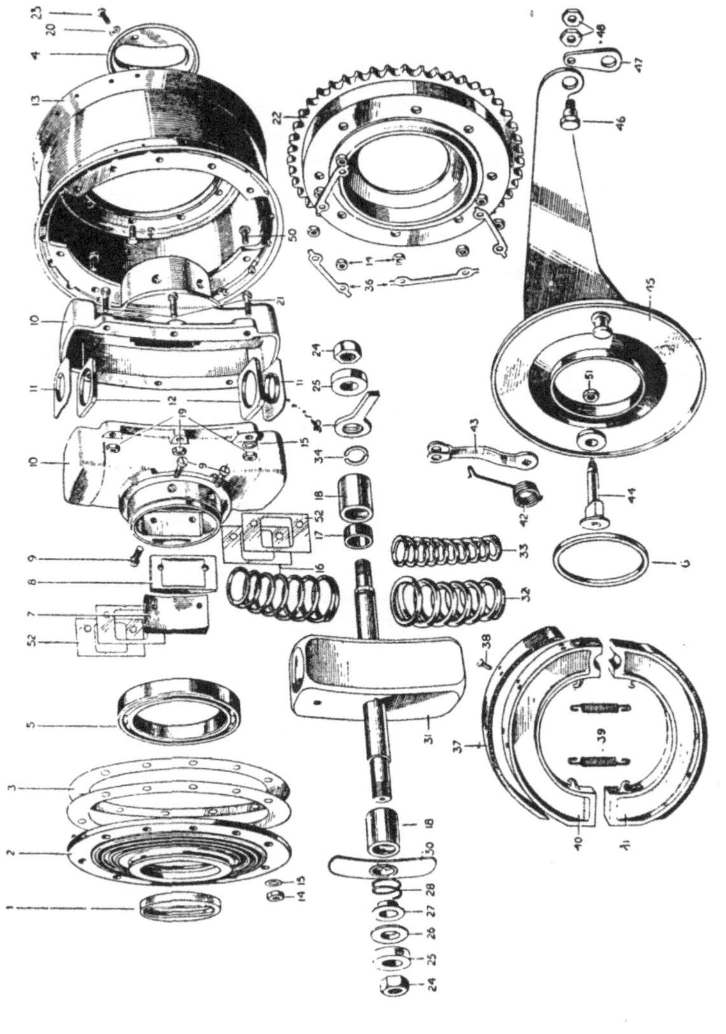

FIG. 51A.—TRIUMPH SPRING WHEEL, MARK II.

KEY TO PART NOS. FIGS. 51 AND 51A (*contd.*).

35. W639
36. W642
37. W631
38. W129
39. W135
40. W618
    W912 (Mk. II)
41. W617
    W911 (Mk. II)
42. W123
43. W74A
44. W856
    W915 (Mk. II)
45. W620
    W853
    W920 (Mk. II)
    W921 (Mk. II)
46. W784
47. W626
    W855
48. W630
49. W641
50. W640
51. S1–52
52. W866
    W867

Spring Wheel Removal and Refitting

(1) Disconnect mudguard rear section.

(2) Remove spring link from rear chain. Engage a gear to avoid chain rolling off the gearbox sprocket.

(3) Unscrew the brake pedal adjuster.

(4) Remove split pin and unscrew nut securing brake anchor plate to brake pedal spindle.

(5) Slacken off wheel nuts and withdraw wheel from frame.

Refitting the Spring Wheel

Procedure is exactly the reverse of removal, but the fitter should ensure that the anchorage lever fits into the channel on the near-side chain-stay lug.

Dismantling Spring Wheel (Mark I)

(1) Remove spindle to frame locator. To remove dust excluder, withdraw distance collars, then centre sleeve, spring and dust plate. To extract dust cover from hub, gently prise it away.

(2) Place wheel spindle in vice and unscrew brake anchor plate in anti-clockwise direction. When unscrewing, either remove brake lever spring or hold lever in such position as to prevent brake shoes binding on brake drum.

(3) Turn wheel over and remove ten nuts securing backplate. Ease plate off studs and carefully remove shims.

(4) Remove spring box assembly from hub. To remove race inner tracks a special puller should be employed.

(5) To remove the outer race from the drive side a suitable piece of hard wood should be employed. The race is removed by driving from the outside of the hub.

(6) The offside outer race which is situated in the back-plate should be removed by a special tool, or the plate heated sufficiently to allow the race to drop out. Great care must be taken if the latter operation is carried out, as excessive heat will destroy hardening of the race.

(7) To dismantle the spring box assembly, remove the six nuts and bolts and tap the cases about $\frac{3}{8}$ in. apart. Now place the assembly in the special jig as illustrated in Fig. 52, with the curvature of the box towards the jig and the long spindle on the left-hand side. Screw down the jig until the springs are sufficiently tensioned to remove the cases.

(8) Unscrew the jig until all spring tension has been removed and withdraw the springs from the shoe. Note that the bottom

of the shoe has two springs fitted, one heavy and one light gauge, the upper part being fitted with one only of medium gauge.

Assembly of Spring Wheel

(1) Place the shoe on the bench with the longer spindle on the left and the convex side uppermost. Insert the heavy and light

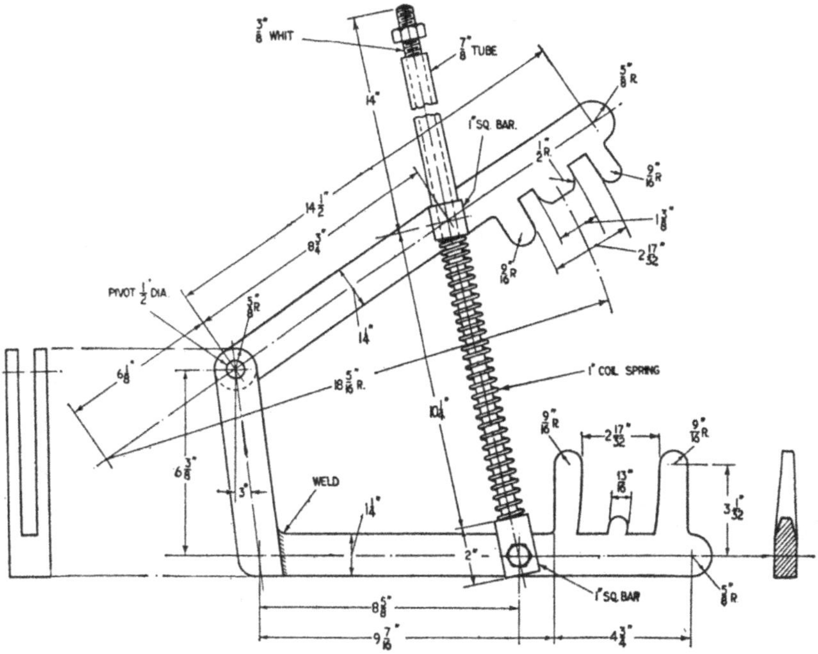

FIG. 52.—DIMENSIONS OF JIG FOR USE WITH THE TRIUMPH SPRING WHEEL.

spring into the bottom hole with the curvature conforming to the shoe. Place the other spring in the top hole in a like manner. Now fit the rubber buffers and spring plates to the springs, plate curve conforming to the shoe, the plate with the smaller hole fitting over the secondary spring.

(2) Fit the assembly to the jig, the two springs being at the bottom. Screw down the jig until the springs have been compressed sufficiently to enable the spring box casing to be fitted. Fit the casing, and unscrew the jig to remove assembly.

(3) Through the gap between the cases, fill with lubricant and tap the cases together with a hide hammer. Fit and tighten up the six nuts and bolts.

(4) Replace the drive side outer race to the hub and ensure that it is fully home.

(5) Fit the inner race to the spring box and grease before assembly. Note the ball cage assembly should be fitted with cage towards the hub centre.

(6) Insert the spring box into the hub.

(7) Assemble the race to the off-side with the balls facing towards the back-plate.

(8) Fit the shims to the hub and assemble back-plate. Screw up the nuts securing the back-plate and test the rotation of the spring box for freedom during the tightening. Failure to do this after new parts have been fitted may result in indentation of the outer ball race or fracture of the hub parts. Any tightness can be obviated by fitting extra shims, allowing the spring box to rotate and at the same time showing no lateral movement.

(9) Place the wheel in the vice or a suitable jig with the drive side uppermost and screw on brake anchor plate. Turn back the anchor plate after fully tightening until the wheel turns freely. The brake cam lever bush should then be lined up with the wheel spindle for assembly purposes.

(10) Fit the dust excluder and then locate the dust plate, ensuring that the spigot is positioned. Place the spring over the spindle and assemble centre sleeve. Fit the spindle to frame locator on the near side and bevel washer to the off side, bevel facing upwards. Where the rubber gaiter is fitted it is advisable to centralise the spindle; here a rig will have to be made to overcome the spring tension.

Maintenance of Spring Wheel

There are two types of spring wheel known as Mark I and Mark II. Method of identification between the two is that the Mark II has raised rings on the back plate and no provision for greasing externally.

Mark I.—Early models had rubber oil-seal gaiters fitted; this type of seal is liable to deterioration, and it was therefore very quickly superseded by the metal mechanical type; modification sets are available, but it is suggested that the work is carried out by your dealer or the wheel returned to the Triumph Service Department. Greasing should be carried out every

1000 miles, one full grease-gun (as supplied in the tool-kit) being sufficient.

Mark II.—This wheel is of a more robust design, but still retains its previous characteristics. The wheel rotates on deep groove ball-journal bearings, and the slipper blocks have been replaced by two case-hardened steel rollers. The previous rubbing movement of the slipper blocks has been converted into a rolling action, thereby practically eliminating wear. As this wheel has no grease-nipples fitted, it is fully charged with graphited grease when assembled and gives an estimated life of at least 20,000 miles before re-greasing is again necessary.

Dismantling Spring Wheel (Mark II)

(1) After the wheel has been removed from the frame take off both spindle nuts and distance collars. Place the wheel on the work-bench with the brake side uppermost.

(2) Prise off the spindle to frame-anchorage lever by using a screwdriver under the lever close to the spindle and then tap the other end of the lever with a hammer. On the underside of the lever are two split collars. Withdraw the dust-excluder centre sleeve, spring and sliding portion. Unscrew the two screws securing dust cover and gently ease the cover away with a screwdriver. The brake anchor plate may be lifted off.

(3) Before removing the slipper roller, test for freedom of rotation and clearance each side with a feeler gauge. If the roller rotates freely and the clearance is not more than 0·002-in. EACH SIDE, it will be in order to carry on with the dismantling. Should the clearance be above the limit make a note of the dimensions to enable rectification on assembly.

(4) Turn the wheel over and remove the chamfered collar and dust-cover assembly. Remove the slipper roller after checking.

(5) Remove the ten nuts securing the back plate to the hub. Gently bump the underside of the spindle on the bench to part the back plate away from the hub. Now ease the back plate off, taking care not to damage the aluminium shims beneath. When the shims have been removed it is advisable to count them to ensure correct clearances on re-assembly.

(6) Withdraw the spring box assembly from the hub.

(7) To remove the brake drum and sprocket, bend back the four locking tabs and remove the eight nuts. Between the brake-drum side bearing and anchor plate a cast-iron ring is fitted.

(8) Removal of the off-side bearing is simplified by heating the back plate; when the temperature is sufficiently high (approx. 100° C.) the bearing can easily be tapped out. On the brake-drum side a suitable piece of hard wood should be made to fit the bearing inner ring in order that it can be drifted out.

Dismantling the Spring Box (Mark II)

(1) Unscrew the six nuts and remove the bolts out of the cases. Tap the cases about ⅜ in. apart, this being sufficient to allow the assembly to be placed in the jig.

(2) Place the spring box in the jig with the curvature towards the jig and the long spindle on the left-hand side. Screw down the jig until the springs are sufficiently tensioned to remove the cases.

(3) Unscrew the jig until the spring tension is released and remove the springs from the shoe. Note that the bottom of the shoe has two springs fitted, one heavy and one light gauge, the upper part being fitted with one only of medium gauge.

(4) If the slipper pads show no sign of wear there is no necessity to disturb them. Wear is generally noticed on the thrust side only, rear on the chain side and front on the opposite. To remove, unscrew the two screws holding each pad in position. Note at the back of each pad metal shims are fitted of varying thicknesses. (See assembly of spring box for re-fitting.)

Reassembling the Spring Box (Mark II)

Lubrication Note.—The ball-journal bearing must be packed with bearing grease only. (See chart on page 211).

(1) If the steel slipper pads have been removed for replacement or adjustment purposes it will be necessary to assemble the spring box less the springs, plates and rubber buffers. This will enable the fitter to operate the axle and guide in the case when testing the roller clearances.

(2) The clearance between the roller and slipper pad must have an equal clearance of 0·002 in. on each side, to allow for any deformation after the springs have been fitted. To enable the fitter to obtain the correct clearance aluminium shims of 0·002 in. and 0·003 in. thickness are available. When the pads have been fitted ensure that the screws securing them to the case are well tightened with a suitable screwdriver.

(3) After the pads have been fitted correctly the case should be parted and then reassembled with the springs. If the above work has not been necessary as per item 3 in the dismantling procedure, assembly can proceed from this point. Place the axle and guide on the bench with the longer side of the axle to the left and the convex side uppermost. Lubricate and insert the heavy and light spring in the bottom hole with the curvature of the spring conforming to the guide. Place the other spring in the top hole in a like manner. Now fit the rubber buffers and spring plates to the springs, plate curve conforming to the guide.

(4) Fit the assembly to the jig, the two springs being in lower position. Screw down the jig until the springs have been compressed sufficiently to enable the spring-box casing to be fitted. Fit the casing, and unscrew the jig to remove assembly.

(5) Through the gap between the cases, fill with lubricant and then tap the cases together with a hide hammer. Fit and tighten up the six nuts and bolts.

Reassembling the Spring Wheel (Mark II)

(1) Fully load the bearings with grease and assemble to the back plate and brake drum. The bearing chip shield should in each case face towards the hub centre.

(2) Fit the bolts to both sides of the hub shell.

(3) Assemble the brake drum to the hub and lock the nuts with new locking plates.

(4) Assemble the spring box to the hub and press well home into the bearing.

(5) Fit the shims to the hub and ensure the correct number as removed have been replaced.

(6) Replace the back plate and tighten up the eight nuts securely.

(7) Grease the slipper pads and spindle on the back-plate side and fit the slipper roller.

(8) Fit the dust excluder, concave facing forward, centre sleeve spring and sliding portion. The chamfered collar should be fitted with the chamfer towards the hub centre.

(9) Turn the wheel over and fit the distance sleeve over the spindle, then grease the slipper pads and fit the slipper roller.

(10) The cast-iron ring should be placed over the spring-box case so that it abuts against the inner ring of the bearing.

(11) Assemble the brake-anchor plate assembly to the brake

drum; it is advisable at this stage to release the spring off the brake lever to prevent the shoes binding in the drum.

(12) Replace the dust-excluder assembly in the same manner as the opposite side, and secure the cover with the two screws. Place the two split collars in the spindle groove and fit the frame-anchorage lever. Re-fit the brake-lever spring and fit the wheel to the frame.

CHAPTER V

FRONT FORKS

Two types of front fork are fitted to Triumph motor cycles—the girder type and the telescopic type. The girder type is dealt with first.

To remove girder-type front forks from the machine :—

(1) Place a box or other suitable article under the crankcase to raise the front wheel off the ground.

(2) Disconnect brake cable at handlebar lever.

(3) Remove headlamp. Where the instrument panel is fitted in the petrol tank, the electric cables should be disconnected at the headlamp. To do this, remove the lamp rim and reflector assembly and pull the cables out of the bulb holder sockets. If the lamp has the panel fitted to the shell as in Fig. 54, remove the three securing screws and withdraw the complete assembly. The panel with wires can be extracted after No. 5 operation has been carried out.

(4) Remove bolt securing top fork-spring anchor.

(5) Remove the rear top and bottom fork spindles.

The fork complete with front wheel mudguard and headlamp can now be removed from the machine for attention. The further dismantling of the fork presents no difficulty.

Steering Head

To dismantle the steering head, first remove the steering damper. Take off the handlebar caps and release the handlebar. Screw off the large adjusting nut at the top of the fork crown after slackening off the pinch bolt and taking off the head clip, removing the crown and stem from below.

Cups and Cones

The bearing tracks on the cups and cones should be carefully examined and if the slightest pitting or damage is observed

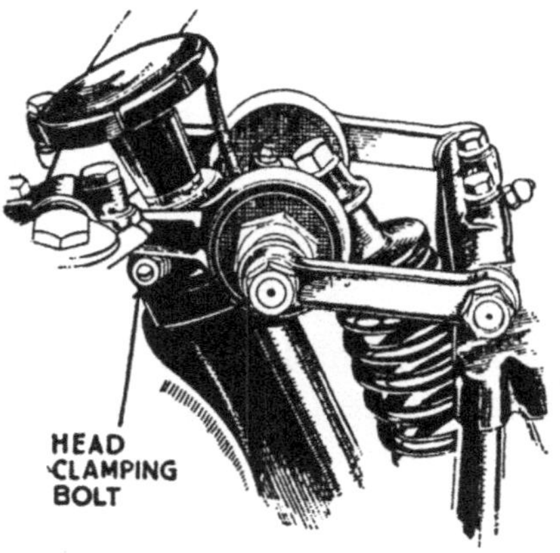

FIG. 53.—THE CLAMPING BOLT AND RUBBER SHOCK ABSORBERS ON THE GIRDER FORK ASSEMBLY.

the part must be changed. It is always desirable to assemble with new bearing balls. No difficulty will be experienced in tapping out the cups from the frame head lug with a suitable drift. Damage to the cups and cones is usually the result of maladjustment of the steering head. If the machine is run with the head slack, hammering will occur.

Spindle Bushes

Worn spindle bushes can be removed by tapping out with a long drift. New bushes should be pressed in, after which they must be cut to size with a line reamer. It is not satisfactory to use a reamer on each bush separately.

Assembly of Girder-type Front Forks

The fork can either be assembled and fitted to the fork crown and head clip or, if these parts have been removed from the machine, the whole can be assembled and fitted to the frame head lug. The

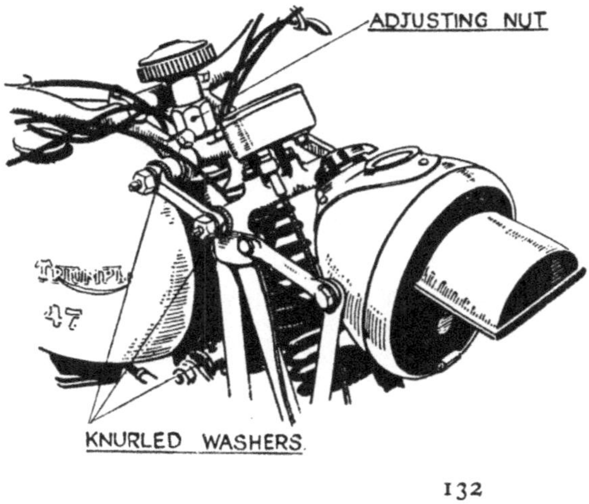

FIG. 54.—THE KNURLED WASHERS ON GIRDER TYPE FORKS WHICH ARE USED FOR TESTING FORK ALIGNMENT.

cups should be filled with grease and the balls pressed into them in order to facilitate assembly. The knurled washers are fitted for testing the adjustment.

Adjusting Fork Links

The knurled washers should be just free to turn, but there should be no side-play. To adjust slacken off bolt end nuts and by means of the squared end turn the fork spindle the estimated correct amount. Tighten up the spindle nuts and test again. Adjustment is best made with the fork fitted to the machine. As each spindle is adjusted the fork should be bounced up and down to make certain that the spindle is free. Carry out this test as each adjustment is made, as otherwise if the fork is not free, each spindle must be slackened off in turn to locate the " offender ".

Adjusting Steering Head

Unscrew the steering damper knob until it is quite free. Slacken off the pinch-bolt nut and gently tighten down the large adjusting nut as far as it will go, but do not apply more than " two finger " pressure on the spanner. Place a box or other suitable article under the crankcase to lift the front wheel off the ground. Slacken off the large adjusting nut a quarter of a turn and test the steering for freedom by turning the forks on to the full lock in both directions. If the steering is not entirely free, the nut should be slackened off a little farther. Finally, tighten up the pinch-bolt nut.

TRIUMPH TELESCOPIC FORK

The Triumph telescopic hydraulically controlled fork will require little attention other than an occasional check of the external nuts, screws and washers. Due to its construction the estimated time between the fork overhauls is about 20,000 miles. At no time during normal service will the forks require

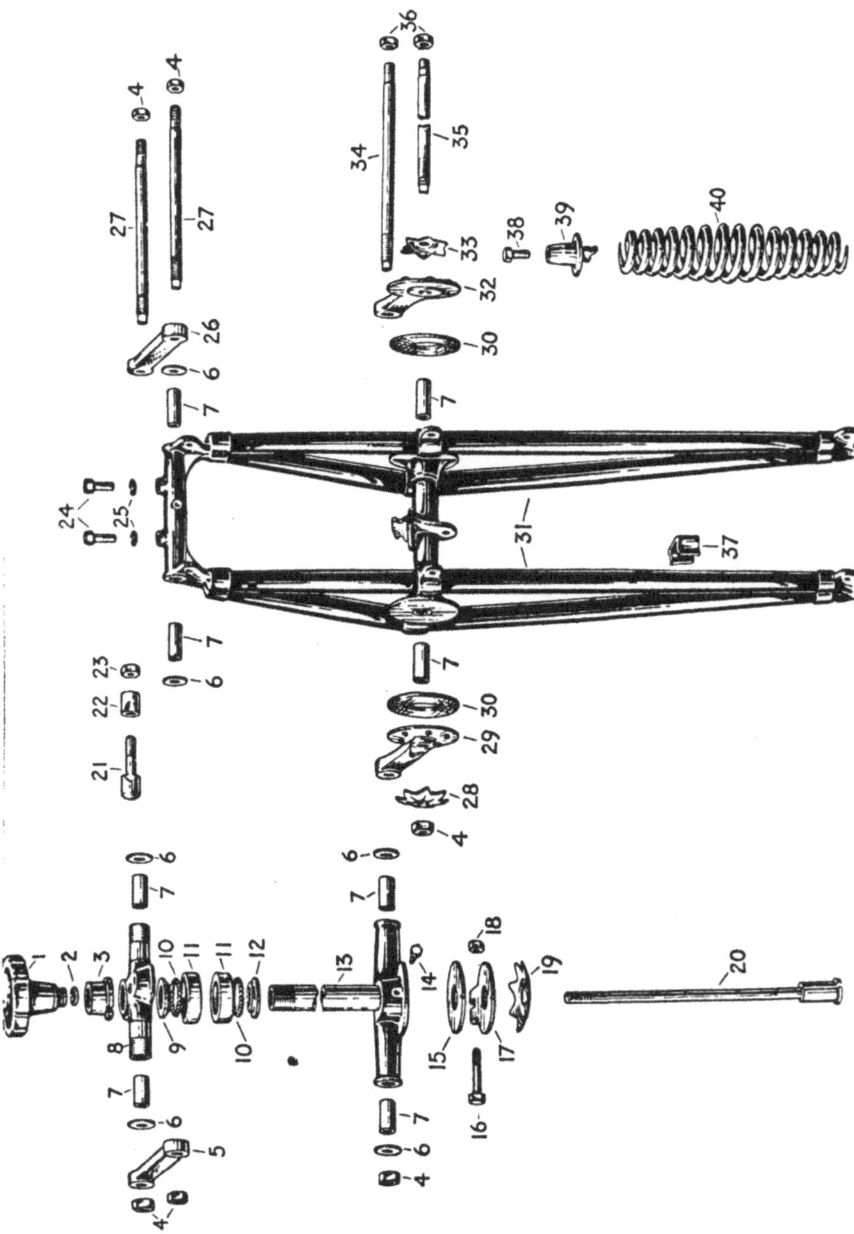

FIG. 55.—FRONT FORKS AND STEERING HEAD ASSEMBLY.

FIG. 56.—TESTING THE HEAD RACE BEARINGS ON A GIRDER FORK MODEL.

topping up with oil; slight leakage that may have taken place will not affect the fork action. Should the leakage of oil have been excessive, it will be necessary to drain the forks completely and re-fill each leg with 100 c.c. (one-sixth of a pint) of oil. (See lubrication chart.) From 1945/8 the fork remains unchanged, but was modified to accommodate the 1949-type headlamp and instrument nacelle. To avoid confusing the fitter each type will be dealt with separately.

KEY TO PART NOS. IN FIG. 55.

1. DS6	8. H177A(WD)	16. F1718	24. DS57	32. H186
2. TS269	9. XS9	17. H182	25. F774E	33. H134
3. H94	10. S70–4	18. S1–51	26. H164	34. H179
4. H193	11. XS10	19. H95	27. H165	35. H180
5. H163	12. XS8	20. H190A	28. H133A	36. H194
6. H191	13. H173	21. TS11	29. H185	38. DS58
7. H168	14. H200	22. TS12	30. DS39	39. H167
8. H177	15. NS167	23. S1–4	31. H183	40. H184

1945/8 Type Forks

To re-fill the forks with oil, the machine should be lifted on to the rear stand, and a box placed under the crankcase high enough to raise the front wheel clear of the ground. This procedure removes all tension from the main spring and enables the operator to unscrew the two large cap nuts situated in the top fork lugs. It will be noted that the nuts after their removal from the head lug are still connected to the hydraulic rod. These parts should not be separated, as the correct measure of oil can be poured into each leg past the spring. Ensure that the drain plugs are quite tight before commencing to re-fill, reassemble the cap nuts to the head lug, remove the box from under the crankcase and test the fork action.

Removing and Dismantling

(1) Raise the machine on to the rear stand, disconnect the front brake at the handlebar.

(2) Lift the front of the machine and place a box of suitable height under the crankcase to give the front wheel ground clearance.

(3) Remove the front wheel; take off the spindle nut on the near-side fork member and slacken off pinch bolt on the off-side fork. Remove brake attachment and brake anchor-plate

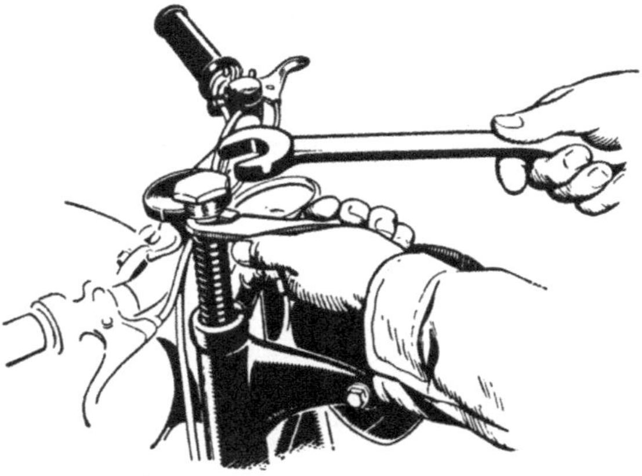

FIG. 57.—METHOD OF REMOVING THE LARGE TOP NUT FROM THE HYDRAULIC ASSEMBLY ROD IN ORDER TO CHANGE THE FORK SPRINGS.

bolt. The spindle can be withdrawn by tapping through from the threaded end and the wheel removed.

(4) Remove front stand and mudguard.

(5) Slacken off the two large cap nuts situated at the top of the forks.

(6) Remove the two bolts situated in the fork spindle lugs; these secure the hydraulic assembly to the lower fork members.

(7) Take out the hydraulic assembly complete by gripping the large cap and lifting.

(8) Slacken off the fork crown pinch bolt and withdraw stanchion and lower tube member assembly.

(9) Remove the dust excluder from the top of the lower member. The stanchion can now be released from the member by pulling sharply; this removes the bush at the same time.

(10) To remove the felt strip from the dust excluder, remove the bottom steel ring, and the felt strip will then be accessible.

(11) At the lower end of the stanchion there is a white metal bearing; to remove this it is necessary to grip the stanchion in a suitable clamp and remove the hydraulic stop nut.

(12) To dismantle the hydraulic assembly compress the main spring and grip the oil-restrictor rod with a pair of pliers. The large cap nut can now be unscrewed off the rod and the spring removed.

(13) Grip the valve body (this is the lower portion of the valve assembly) and unscrew the top support body. Remove the restrictor rod from the pressure tube.

(14) To remove the restrictor valve, unscrew the nut and the restrictor can be pulled off the rod with the restrictor cup. Withdraw the cup-pin from the rod.

(15) Remove the top dust-cover tube which still remains in position and extract the felt washer oil seal. (To dismantle steering head, see page 142.)

Inspection of Telescopic Forks

With the fork completely dismantled all parts should be thoroughly washed and laid out for inspection. Examine both upper and lower tube members for dents or distortion. The stanchion must also be checked: this should be placed between lathe centres and spun. If a dial test indicator is placed at various intervals along the stanchion whilst the latter is still in the lathe, a true reading will be given of any inaccuracy. Check the main spring free length from the data

KEY TO PARTS.

1. H405 (500 c.c.) H404 (350 c.c.)
2. H380
3. H432
4. H384 (500 c.c.) H505 (Sidecar) H385 (350 c.c.)
5. H422
6. E667A
7. H518
8. H382
9. H390
10. H415
11. H428
12. H519
13. H430
14. F1760
15. H379
16. H414
17. H441
18. H431
19. H391
20. H445
21. H420
22. H421
23. H408
24. H409
25. H95
26. F1718
27. S25-3
28. S1-51
29. H413
30. H435
31. H402
32. H439 (500 c.c.) XS8 (350 c.c.)
33. H438 (500 c.c.) H433 (350 c.c.)
34. H407 (500 c.c.) H406 (350 c.c.)
35. H450
36. H379
37. F1760
38. H506
39. H436
40. DS6
41. H397
42. H399
43. H454
44. H453
45. H452
46. F879
47. H394
48. H395
49. E1335
50. H398
51. H443
52. H444

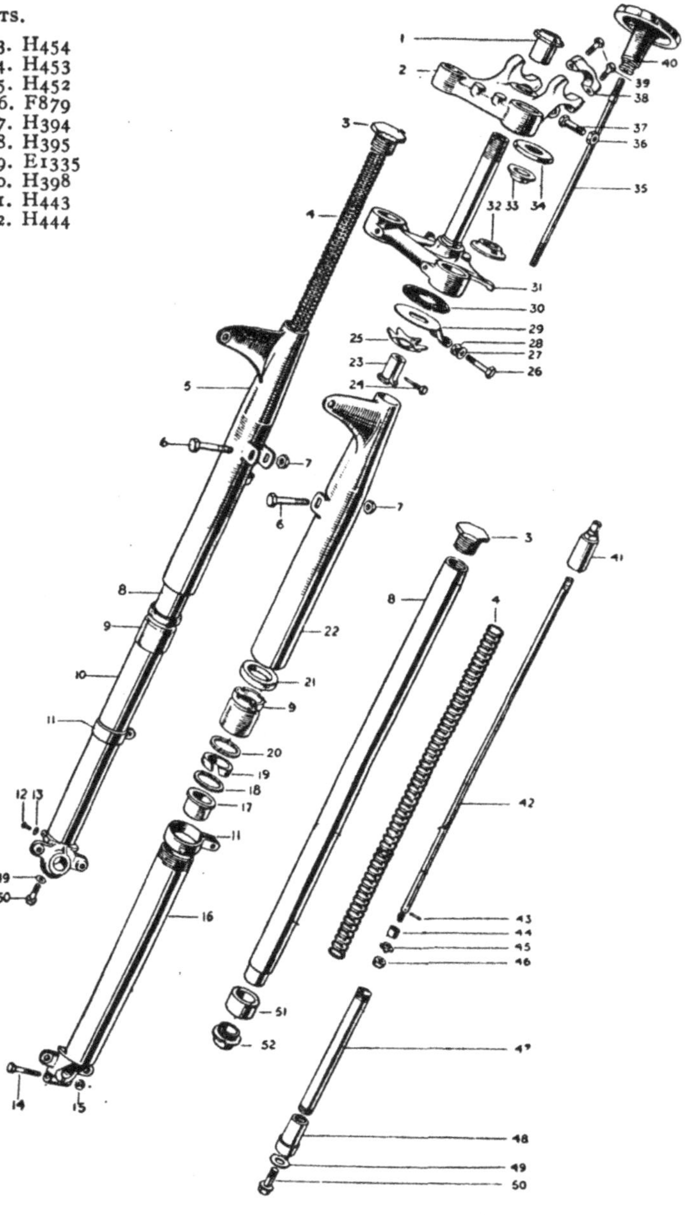

FIG. 58.—COMPONENTS OF TRIUMPH HYDRAULIC FORK (1ST CONDITION).

KEY TO PART NOS.

1. H405
   H404
2. H663A
3. H432A
4. H384
   H505
   H385
5. H657
6. H570
7. H518
8. H382A
9. H390
10. H415
11. H428
12. H519
13. H430
14. F1760
15. H379
16. H414
17. H441
18. H431
19. H391
20. H445
21. H420
22. H656
23. H408
24. H409
25. H95
26. F1718
27. S25-3
28. S1-51
29. H413
30. H435
31. H695
32. H439
    XS8
33. H438
    H433
34. H407
    H406
36. H379
37. F1760
38. S1-51
39. H660
40. H536
41. H397
42. H399A
43. H454
44. H453
45. H452
46. F879
47. H394
48. H395
49. E1335
50. H398
51. H443
52. H444
53. H565
54. H618
55. H519A
56. H624
57. H623

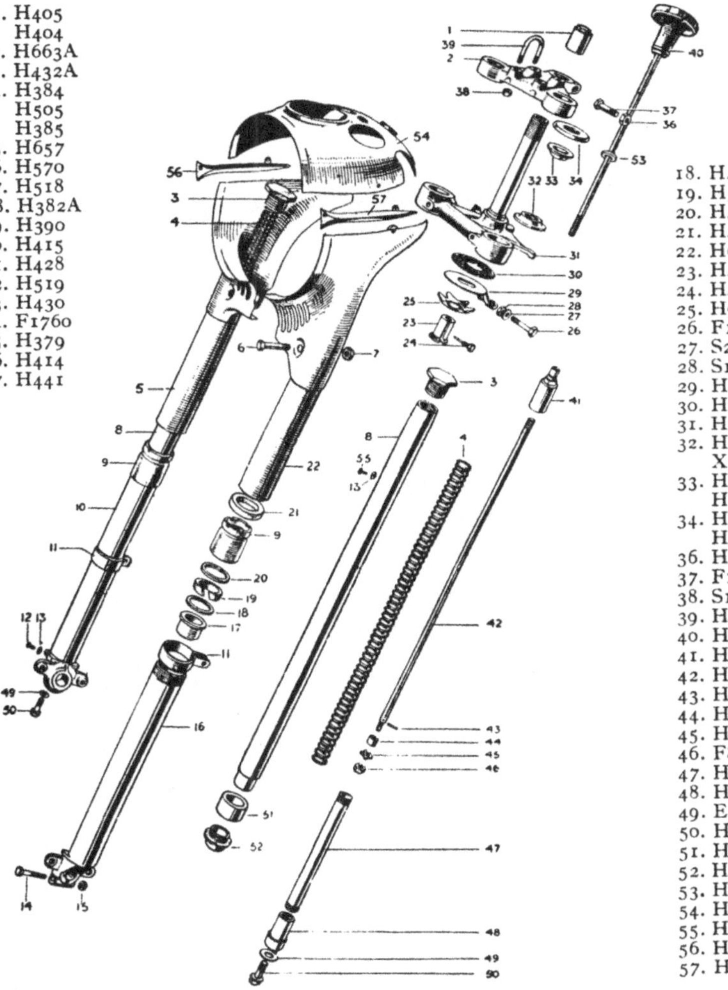

FIG. 58A.—TRIUMPH TELESCOPIC FORK (2ND CONDITION).

table, page 202. As the fork will have a further life of 20,000 miles after overhaul, it is desirable that the following parts are fitted to ensure a trouble-free mileage :—

2—H 443.	Stanchion lower bearing.
2—H 441.	Stanchion upper bearing.
2—E 1335.	Copper washer.
2—H 452.	Oil restrictor.
2—H 453.	Cup for oil restrictor.
2—H 454.	Pin securing oil restrictor cup.
2—H 391.	Dust excluder felt.
2—H 420.	Top cover felt washer.
2—H 430.	Drain-plug washer.

Having prepared the fork and acquired the new parts, the fork should be assembled in the following manner.

Assembly

Lubricate all parts before assembly.

(1) Fit the white-metalled bush to the stanchion and lock with the hydraulic stop nut. Check bush for freedom of movement.

(2) Assemble the drain plug and washer to lower member, ensuring its tightness, otherwise complete loss of oil may result.

(3) Fit the new felt washer to the dust excluder, first by fitting the thin metal washer, then the felt and finally the outer washer.

(4) Assemble the stanchion to the lower member and fit the top bearing. Screw on the dust excluder and check the movement of the stanchion and the lower member ; this should be free and smooth.

(5) To assemble the hydraulic valve assembly, fit the pin to the restrictor rod, then the cup and restrictor and lock the assembly with the nut. Screw the lower valve body to the pressure tube ; insert the restrictor rod assembly and screw the top support sleeve to the pressure tube. To ensure correct operation of the valve, place the assembly in a tin of oil and pump the rod up and down when the pressure tube is filled ; the upward movement of the rod should be restricted and the downward movement unrestricted.

(6) Fit the top cover tube to the fork crown, insert the pinch bolts but do not tighten. Place the felt washer into the cover tube and slide the stanchion with lower member through the

felt washer and fork crown until the taper on the stanchion engages with the fork head lug. Using the large cap nuts tighten both stanchions to the head lug and then tighten up the pinch bolts in the fork crown.

(7) Remove the cap nuts from the stanchion.

(8) Fit the main spring to restrictor and depress sufficiently to enable the rod to be gripped with a pair of pliers, to aid the assembly of the cap nut to restrictor rod.

(9) Insert the hydraulic assembly into the stanchion. It will be noted that a dowel is fitted into the body of the pressure tube ; this must be located in the dowel hole in the lower member. Fit the support bolt with copper washer through the spindle lug and screw into hydraulic assembly, lock tight.

(10) Push the lower tube member up into the upper cover and pour 100 c.c. (one-sixth of a pint) of oil into the top of each stanchion. Now force the leg down by depressing the cap nut. Screw the cap nut into the stanchion and ensure that it is well tightened. If this precaution is not taken the stanchion taper will not be drawn into the fork head lug and excessive strain will be put on to the crown and stem.

(11) Fit the front mudguard and front stand.

(12) Fit the front wheel, but do not tighten the spindle nut or pinch bolt.

(13) Connect up lower brake attachments and brake anchor plate to mudguard clip.

(14) Remove the box from under the crankcase. See instructions regarding wheel alignment on page 119.

Changing Main Springs

In order to change the main springs and fit stronger ones for sidecar purposes, the front wheel should just be raised as described in paragraphs (1) and (2) dismantling procedure. Remove the two cap nuts in the fork head lug. Remove the box from under the crankcase; this will make the springs more accessible. Grip the spring and force it down until the restrictor rod can be gripped with a pair of pliers. The cap nut can now be unscrewed off the restrictor rod, and the spring removed. Repeat this operation on the opposite fork leg. Position the new fork springs and reassemble the forks in exactly the reverse procedure. The fork springs have a colour identification as follows :—

3T	(solo)	white
5T, T100 and 6T	(solo)	red
,, ,, ,,	(sidecar)	blue
TR5	(solo)	green

Steering Head

To dismantle the steering head after removal of the forks, first remove the steering damper. Take off the handlebar caps and release the handlebar. Screw off the large adjusting nut at the top of the fork head lug after slackening off the pinch bolt and take off the head-lug clip. Remove the crown and stem from below.

Cups and Cones

The bearing tracks on the cups and cones should be carefully examined and if the slightest pitting or damage is observed the parts must be changed. It is also desirable to assemble with new bearing balls. No difficulty will be experienced in tapping out the cups from the frame head lug with a suitable drift. Damage to the cups and cones is usually the result of maladjustment of the steering head. If the machine is run with the head slack, hammering will occur.

The cups should be filled with grease and the balls pressed into them in order to facilitate assembly. The 350-c.c. Twins have twenty-two balls top and bottom and the 500-c.c. have twenty-two balls top and twenty balls bottom. Assemble the fork crown and stem to the head, place the head lug in position and screw down the sleeve nut. Fit the pinch bolt, but do not finally tighten until the telescopic fork legs are positioned. Replace the handlebars and steering damper assembly.

Adjusting Steering Head

The adjustment is the same as for the girder fork, as described on page 133. Test at the same time the adjustment of the steering head races by grasping the lower end of the

FIG. 59.—TESTING THE PLAY ON THE FORK BUSHES OF THE TELESCOPIC FORK.

The adjustment of the steering head should be checked by watching the head assembly while the forks are moving backwards and forwards.

front forks and alternately pushing and pulling (see Fig. 59). No play should be felt.

Crash Repairs

Minor damage repairs with this type of fork are permissible, but any repairs requiring extensive operations to straighten the parts should definitely not be undertaken. The material may be structurally damaged, causing disastrous results when the forks are again subjected to road conditions. To test the alignment of the stanchion, this part should be placed in between lathe centres. The stanchion should now be spun, and with the aid of a D.T.I. (dial test indicator) the highest reading will be found. In the absence of a lathe, the stanchion can be rolled on a surface table or a piece of plate glass to ascertain the amount of bow. The maximum amount should not exceed $\frac{5}{16}$ in. if the stanchion is to be put into further service after straightening. To straighten, place the stanchion between two swage blocks of suitable radius and apply pressure to the highest point by means of a press. Remove the stanchion and carry out the tests mentioned earlier. It will probably be necessary to carry out this procedure three or four times before the stanchion is completely true. In no circumstances must the stanchion be used if there is any sign of kinking or the damage has taken place in a very small area.

The top head lug and fork crown and stem lug can be restraightened to their original position in the cold state only if the damage is slight. Extensive adjustment of the lug will cause fracture or weakening of the structure. To straighten the lugs a bar of suitable size should be placed through the offending stanchion hole. Then with the part clamped in the vice apply pressure on the bar in the direction needed. When carrying out this procedure on the lower lug ensure that the pinch bolts are securely tightened up. The lugs must now be suitably rigged to enable a thorough check to be made. If the operator has no suitable jig to enable him to carry out this check, the fork crown and stem and top lug can be assembled to the frame and the stanchions positioned and bolted up. Any sign of twist or the stanchion not freely entering the taper in the top lug will be quite obvious.

1949/51 Type Forks

Mechanically this fork is the same as the 1945/8 types described on page 133 with the exception of the filling procedure. To refill the forks with oil, remove the headlamp rim assembly from the nacelle which will give access to the upper part of the stanchions. In each stanchion a screwed plug is fitted; when these plugs have been removed the correct amount of oil can be introduced into each fork leg by means of a pressure oil can or gun through the headlamp aperture. The exception to this is the TR5 fork which remains as 1945/8 pattern.

Removing and Dismantling

(1) Remove headlamp rim assembly and disconnect wires.

(2) Unscrew cable at speedometer head and disconnect light adaptor.

(3) Unscrew steering damper knob and withdraw distance washer.

(4) Take out the three screws and nuts securing headlamp mounting ring and detach ring from nacelle.

(5) Unscrew the six screws securing nacelle top and remove top complete with instruments after disconnecting horn wire and detaching the lighting wires at the plug.

(6) Detach handlebar by removing the two U-bolts.

(7) Raise front wheel clear of ground by placing a suitable box under crankcase.

(8) Remove front wheel, take off spindle nut on nearside fork member and slacken off pinch bolt on offside fork. Remove brake attachment and brake anchor-plate bolt. Withdraw spindle by tapping through from threaded end to release wheel.

(9) Unscrew the two large cap nuts situated at top of forks. At this point it is advisable to remove the two drain plugs in the lower fork members and drain the oil. To quicken the draining, pump lower members up and down.

(10) Slacken fork stem pinch bolt and unscrew sleeve nut.

(11) Tap the top lug away from the taper of the stanchions.

(12) Take out damper bolt, drop fork stem through frame lug and withdraw fork assembly. Take care not to lose the steering balls.

If the mechanic does not wish to disturb the steering column, carry out operations 1 to 9, then slacken off the middle lug pinch bolts. Unscrew the two large cap nuts a few turns, then give each a sharp blow with a hide or lead hammer to break the taper fit between top lug and stanchion. Completely unscrew the cap nuts, remove the two hydraulic assembly bolts situated in the bottom member lugs and withdraw the spring and hydraulic assemblies. Remove fork legs by pulling through the middle lug.

(13) Unscrew the two hydraulic assembly securing bolts in lower member lugs and withdraw spring and hydraulic assembly. Take away top head lug.

(14) Unscrew middle lug to stanchion pinch bolts and remove lower nacelle covers.

(15) Remove dust excluder from top of lower member. The stanchion can now be released from the member by pulling sharply; this removes the bush at the same time.

(16) To remove felt strip from dust excluder, remove bottom steel ring, and the felt strip will then be accessible.

(17) At the lower end of the stanchion there is a bearing; to remove this it is necessary to grip the stanchion in a suitable clamp and remove the hydraulic stop nut.

(18) To dismantle the hydraulic assembly, compress the main spring and grip the oil restrictor rod with a pair of pliers. The large cap nut can now be unscrewed off the rod and the spring removed.

(19) Grip the valve body which is the lower portion of the valve assembly, and unscrew the top support body. Remove restrictor rod from pressure tube.

(20) To remove restrictor valve, unscrew the nut and the restrictor can be pulled off the rod with the restrictor cup. Withdraw the cup-pin from the rod.

For inspection and crash repairs, see pages 137 and 143.

Assembling 1949/51 Forks

Lubricate all parts before assembly.

(1) Fit bearings to stanchion and lock with hydraulic stop nut. Check bearings for freedom of movement.

(2) Assemble drain plug and washer to lower member, ensuring tightness, otherwise complete loss of oil may result.

(3) Fit new felt washer to dust excluder, first fitting the thin metal washer, then the felt and finally the outer washer.

(4) Assemble stanchion to lower member and fit top bearing. Screw on dust excluder and check movement of stanchion in lower member, which should be free and smooth.

(5) To assemble hydraulic valve assembly, fit pin to the restrictor rod, then cup, restrictor, and lock assembly with nut. Screw lower valve body to pressure tube; insert restrictor rod assembly and screw top support sleeve to pressure tube. To ensure correct operation of the valve, place assembly in a tin of oil and pump rod up and down; when pressure tube is filled, the upward movement of the rod should be restricted and the downward movement unrestricted. Assemble spring to hydraulic assembly. Compress spring and grip rod with pair of heavy pliers or in the corner of vice jaw and screw cap nut in position.

(6) If the steering column assembly has been dismantled, first grease the cups in the frame and press the balls into them.

3T and TR5 . .	Size $\frac{3}{16}$ in.	Top 22	Bottom 22
5T, 6T and T100 .	Size $\frac{3}{16}$ in.	Top 22	——
5T, 6T and T100 .	Size $\frac{1}{4}$ in.	——	Bottom 20

Assemble fork crown stem and top lug, tighten down sleeve nut until steering moves freely from side to side with no up and down movement. Fit and tighten pinch bolt and lower damper parts.

(7) Fit the horn.

(8) Assemble lower nacelle covers and position pinch bolts and nuts only.

(9) Place felt washers in nacelle covers and slide stanchions through fork crown lug and into tapers of top lug. Lightly tighten up fork crown pinch bolts.

(10) Insert hydraulic assembly into stanchion. In the body of the pressure tube a dowel is fitted; this must be located in the dowel hole in the lower member. Fit support bolt with copper

washer through spindle lug and screw into hydraulic assembly; lock tight.

(11) Push lower tube member up into upper cover and pour 100 c.c. (1/6 pint) of oil into top of each stanchion. Now force leg down by depressing cap nut. Screw cap nut into stanchion and ensure that it is well tightened. If this precaution is not taken the stanchion taper will not be drawn into the top lug and excessive strain will be put on to the crown and stem. Tighten up fork crown pinch bolts.

(12) Fit handlebars to top lug and tighten up U-bolts. Connect all cables.

(13) Connect up wires from nacelle top to plug by means of matching colours on wires and connect horn wire.

(14) Fit nacelle top, ensuring that control cables and dipper wire are threaded through rubber grommets and that the chrome "flashes" are fitted under the first two of the three retaining screws on each side.

(15) Connect speedometer drive and light adaptor.

(16) Replace steering damper knob, threading washer on stem on underside of nacelle.

(17) Fit lamp front adaptor rim to front of nacelle by three retaining screws; also lamp front retaining hook and spring.

(18) Connect lighting wires to terminals on rear of reflector (earth wire terminal fits on focussing clip screw).

(19) Fit reflector followed by rubber retaining ring and lamp front.

(20) Fit the front mudguard and front stand.

(21) Fit the front wheel.

Changing Main Springs (1949/51 Models)

In order to change the main springs or to fit stronger ones for sidecar purposes, carry out fork removal operations (1), (2), (3), (4), (5), (6) and (7), then screw off the two cap nuts in the fork head lug. Remove the box from under the crankcase; this will make the springs more accessible. Grip the spring and force it down until the restrictor rod can be gripped with a pair of pliers. The cap nut can now be unscrewed off the restrictor rod, and the spring removed. Repeat this operation on the opposite fork leg. Position the new fork springs and reassemble the forks in exactly the reverse procedure. The fork springs have a colour identification as follows :—

3T	(solo)	white
5T, T100 and 6T	(solo)	red
,, ,, ,,	(sidecar)	blue
TR5	(solo)	green

Petrol Tanks

The petrol tank is fitted to the frame with four bolts which each pass through two rubber pads, one on each side of the supporting bracket. These pads are fitted in order to give the tank a resilient mounting; a tank bolted direct to the frame would very soon leak. When refitting a tank it is extremely important that the circular rubber pads are properly positioned and in good condition. Shoulder bolts are fitted, and these should be tightened right up and then secured by means of wire passed through the hole in the head. The front and rear pair of bolts are wired together separately. To remove the petrol tank it is necessary to disconnect the petrol feed pipe and the U-pipe (not fitted to T.100 or post-war 5T) at the front of the tank after draining. The opportunity should be taken to clean the U-pipe, as after a considerable mileage a certain amount of foreign matter will be found to collect in it. It should be blown out with the tyre pump after being allowed to dry out if time permits. Where an instrument panel is fitted it will be necessary to remove this and also to disconnect the oil-pipe, lighting switch and ammeter wires.

CHAPTER VI

CARBURETTER AND CONTROLS

To remove the carburetter for examination proceed as follows :—

(1) Turn off the petrol and remove the petrol pipe.

(2) Where a magdyno is fitted, slacken off the carburetter-fixing nuts and slide back the carburetter. Unscrew the holding bolt under the mixing chamber and remove the float chamber. The fixing nuts can now be removed and the mixing chamber complete with slides withdrawn from the studs. Models fitted with magneto only should first have the holding bolt slackened off and fixing nuts removed. Push the float chamber forward and remove carburetter complete.

(3) Unscrew the mixing chamber cap and withdraw the throttle and air slide.

Carburetter Cleaning and Examination

Completely dismantle the carburetter and thoroughly clean all parts. Remember to slacken off the float chamber cover lock screw before removing the cover. It is desirable to allow the carburetter to dry out, if possible, before cleaning, as this facilitates the removal of any deposit. Examine the float needle for wear and renew if necessary. The needle jet should also be carefully inspected and replaced if worn. A worn needle jet is frequently the cause of heavy petrol consumption. Note that the needle never wears. Wear on the throttle slide is unusual except after long service; where a machine is used under adverse conditions slide wear may be more rapid. When renewing the slide see that it is the correct size; the size and cut-away is marked on it. See data table for further information.

It is always advisable to check over the flange face, as this tends to distort due mainly to uneven tightening or an incorrect flange washer of too great a thickness. To test the flange for bow, place a straight-edge across the face. Where the bow is not excessive it can be removed by rubbing the face on a piece of emery cloth fitted to a flat surface. If the bow cannot be removed in this way, file the face with a 6-in. flat smooth file and finish off as stated above.

When reassembling the carburetter see that all washers are in good condition. Do not tighten fully home the holding bolt, as the float chamber position is best adjusted after the carburetter has been fitted to the engine. Make certain that the clip on the float is engaged in the groove in the needle. It is not normally necessary to remove the slides unless they are to be renewed. Removal is effected by removing the split pin and sliding the cable nipple out of its housing. On some post-war models the throttle slide is attached to the wire by a screwed nipple. On machines dating from 1945 to 1948, the air slide cannot be removed, as this is a component part of the operating plunger mechanism. On earlier and later machines control is by Bowden cable and lever. If the needle is removed, see that it is fitted into the correct position when it is replaced. The normal position is in the second groove from the top.

Refitting Carburetter

To replace carburetter proceed as follows :—

(1) Fit the slides into the body, being careful to insert the needle into the needle jet. There is a positioning projection on the mixing chamber top which fits into a slot in the top edge of the mixing chamber. Secure by screwing down the cap.

(2) Fit the carburetter flange to the two studs on the cylinder barrel induction pipe, using a new washer to make the joint. Tighten up the nuts evenly, a little at a time to avoid distorting the flange.

(3) Position float chamber and tighten up holding bolt.

(4) Fit the petrol pipe and turn on the petrol.

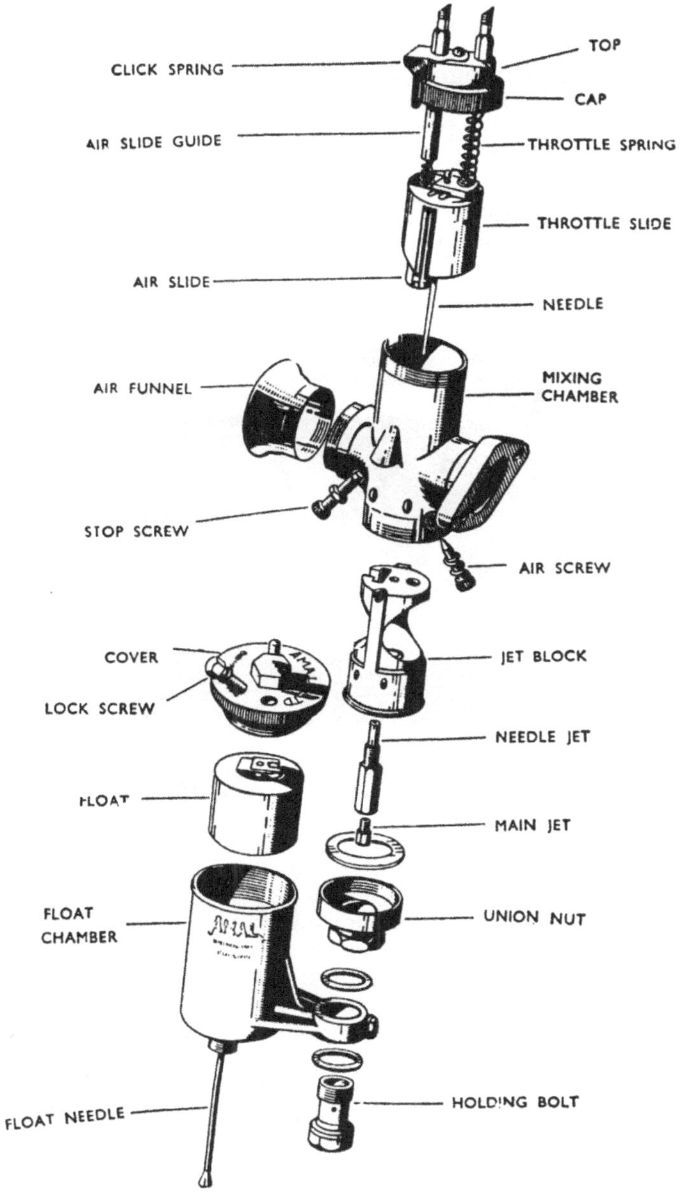

FIG. 60.—THE PARTS OF THE CARBURETTER.

1945/8 models are fitted with an air slide operated by a lever instead of a Bowden cable.

(5) Start up the engine and warm it up a little.

(6) Adjust the slow running by positioning the throttle stop screw so that the throttle is slightly open when the twist grip is closed down. Adjust the slow running mixture strength with the pilot air-adjusting screw. It is best to start with the throttle stop so adjusted that the engine runs rather fast and the pilot air-screw screwed well in, thus giving a rich mixture. This will cause the engine to " hunt "—run alternately fast and slow.

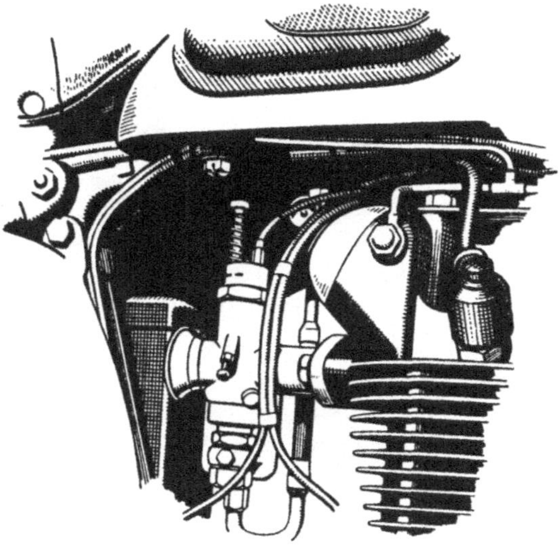

FIG. 61.—THE AIR SLIDE OPERATING LEVER ON THE MODEL 350-C.C. TWIN.

The engine should then be slowed down by screwing out the stop screw and the mixture adjusted with the pilot air-screw until the engine runs evenly and smoothly, with the ignition slightly retarded. Machines with the automatic advance and retard mechanism will have the magneto in the fully-retard position when the engine is running slowly. Do not make this adjustment with the engine cold or fully warmed up. Allow the engine to run for about two minutes from cold before making the slow-running adjustment.

Air-Valve Slide

It is important that the air slide opens fully. If the air slide is opening only partially the mixture will be too rich and the consumption will be heavy. Adjustment can be made on the cable adjuster at the top of the carburetter. Carburetters fitted with the plunger-type air-valve slide will require little or no attention. The valve is held in the open position by means of a spring and when the plunger is depressed and turned in a clockwise direction the slide will be locked in the closed position.

Mixture

The machine is turned out from the Works with the correct main jet and throttle slide. The sizes of these parts have been arrived at after experiment, and no useful purpose will be served by fitting a jet or slide of a different size from the standard part. The needle position is variable, and controls a wide range of throttle opening. The lower the needle the leaner the mixture. Too rich a mixture is indicated by black smoke from the exhaust, "lumpy" running and a sparking-plug the electrodes of which are covered with soot. A lean mixture causes the engine to run hot with some loss of performance; the sparking-plug has a grey, overheated appearance and the running of the machine is improved if the air control is partially closed. Normally the needle should be fitted with the clip in the second groove from the top. If the mixture is still too rich with the needle in the top groove, this is an indication that the needle jet is worn and requires renewal. Do not confuse black smoke from the exhaust with blue smoke caused by worn piston-rings or cylinder bore or both.

Carburetter Faults

Dirt or water will cause engine to stop or only run on small throttle opening. Uneven tightening of the mixing chamber will cause distortion and sticking throttle slide. Flooding carburetter can be caused by float needle not seating, needle bent or punctured float.

First check petrol flow when investigating weak mixture. Starvation at the carburetter can be due to a blocked feed pipe or obstructed tap filter.

Control Cables

The throttle and air cables require no attention with the exception of periodical lubrication. If it is found that a cable is not operating freely in the casing, this is usually

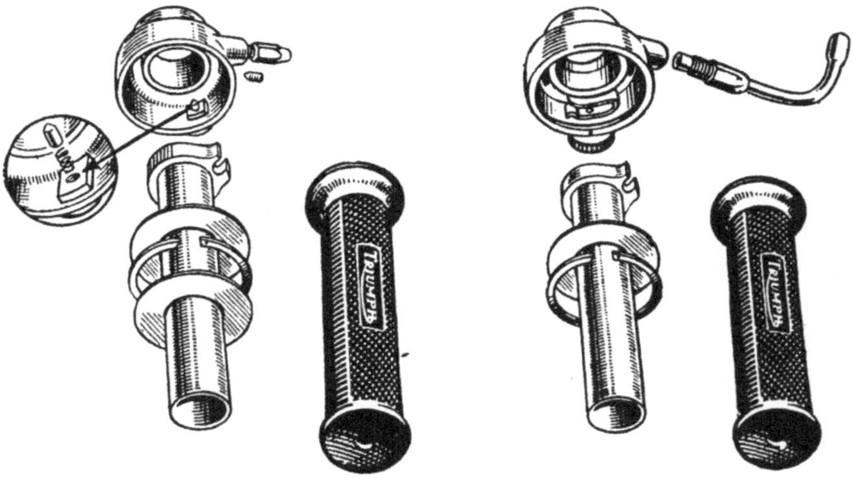

(1ST CONDITION). (2ND CONDITION).

FIG. 62.—TRIUMPH PATENT TWIST.

because it is dry. To lubricate, the cable should be detached from the handlebar lever and the end supported as high as possible. A small funnel should be made with thick brown paper and tied round the end of the casing. This should then be filled with thin oil and allowed to stand overnight, when the oil will penetrate down the casing and effectively lubricate the parts.

Twist Grip Throttle Control

There are two kinds of twist grip in use: ratchet and friction, the latter superseding the former. To dismantle either type, peel back the rubber grip and extract the spring ring and washer. Unscrew the cable stop and withdraw the cable from the twist grip. The grip rotor can then be drawn off, care being taken not to lose the small plunger and spring which provide the click action in the ratchet type. The friction twist grip will require little or no attention, and if the rotor is smeared lightly with grease and replaced it will remain in service for a considerable period.

When re-inserting the cable the grip must be revolved so that it picks up the cable nipple and draws it into place. If the ratchet twist grip is harsh the plunger should be removed and the edge rubbed down on an oil-stone so that it is slightly radiused. When the twist grip will not hold its position the opposite procedure should be carried out by rubbing the flats of the plunger down until the worn edge again takes the correct shape. The plunger should not be chisel sharp; the edge should be slightly rounded.

Air-Filter

The air-filter as fitted to 1949 models can be adapted to fit all twin models from 1937 onwards.

Clean the filter element every 2,000 miles. In extremely dusty conditions service at more frequent intervals. To clean remove the filter lid with element attached and wash in petrol or similar solvent. Allow to dry, and re-oil with Vokes Trifiltrene Oil, or any SAE 20 rating oil.

Maximum power output of the 3T and 5T with the air-filter in use is little affected. With the T.100 it is necessary to disconnect the filter and substitute a 160 main jet for the 150.

CHAPTER VII

B.T.H. IGNITION

The B.T.H. magneto has no manual control over the advancement and retarding of the ignition spark. The ignition timing is controlled by the auto advance mechanism, which varies the advancement of the spark in relation to the engine speed.

Dismantling the Magneto

(1) Remove the contact-breaker cover by releasing the three screws, taking care not to damage the sealing washer situated between the inner face of the cover and the outer face of the contact end-plates; also do not adjust unnecessarily the tension of the earthing plate, which makes contact with the face of the contact-breaker fixing screw.

(2) Remove the hexagon-headed screw in the centre of the contact breaker by means of the contact spanner provided with each machine, when the contact breaker can be withdrawn bodily.

(3) The collector mouldings should next be removed by releasing the knurled retaining nut, care being taken in this instance not to fracture the carbon brushes.

(4) Remove the two safety gap screws and locking tabs situated at the top and bottom of the main housing at the driving end; this is important, as failure to observe this will result in damage to the slip-ring on the armature when the armature is withdrawn from the main housing.

(5) The contact end-plate may now be removed by unscrewing the four fixing screws held in position by special locking washers which must be preserved for reassembling. These washers are specially designed to ensure tightness of the fixing screws against vibration.

(6) Withdraw the armature from the main housing. Owners are strongly advised not to attempt to dismantle the armature for any purpose, as special extractors, etc., are required for this operation.

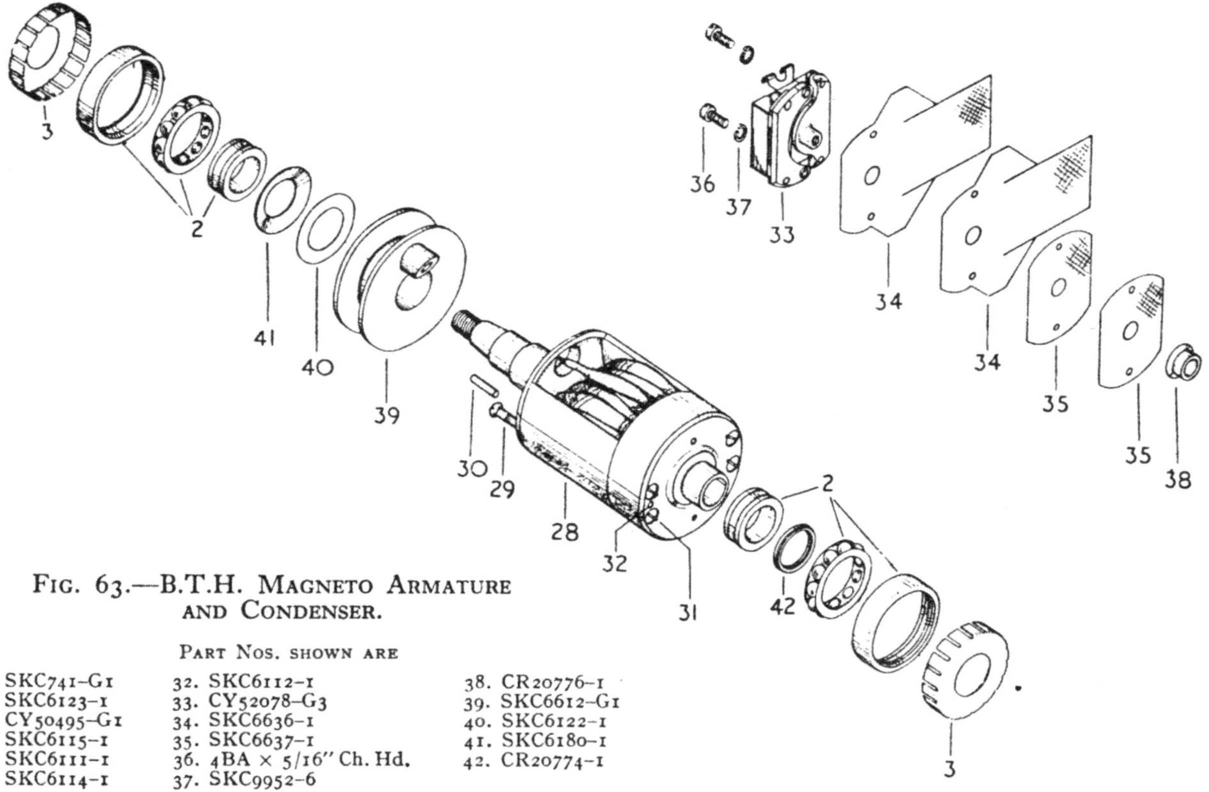

FIG. 63.—B.T.H. MAGNETO ARMATURE AND CONDENSER.

PART NOS. SHOWN ARE

2. SKC741–G1	32. SKC6112–1	38. CR20776–1
3. SKC6123–1	33. CY52078–G3	39. SKC6612–G1
28. CY50495–G1	34. SKC6636–1	40. SKC6122–1
29. SKC6115–1	35. SKC6637–1	41. SKC6180–1
30. SKC6111–1	36. 4BA × 5/16″ Ch. Hd.	42. CR20774–1
31. SKC6114–1	37. SKC9952–6	

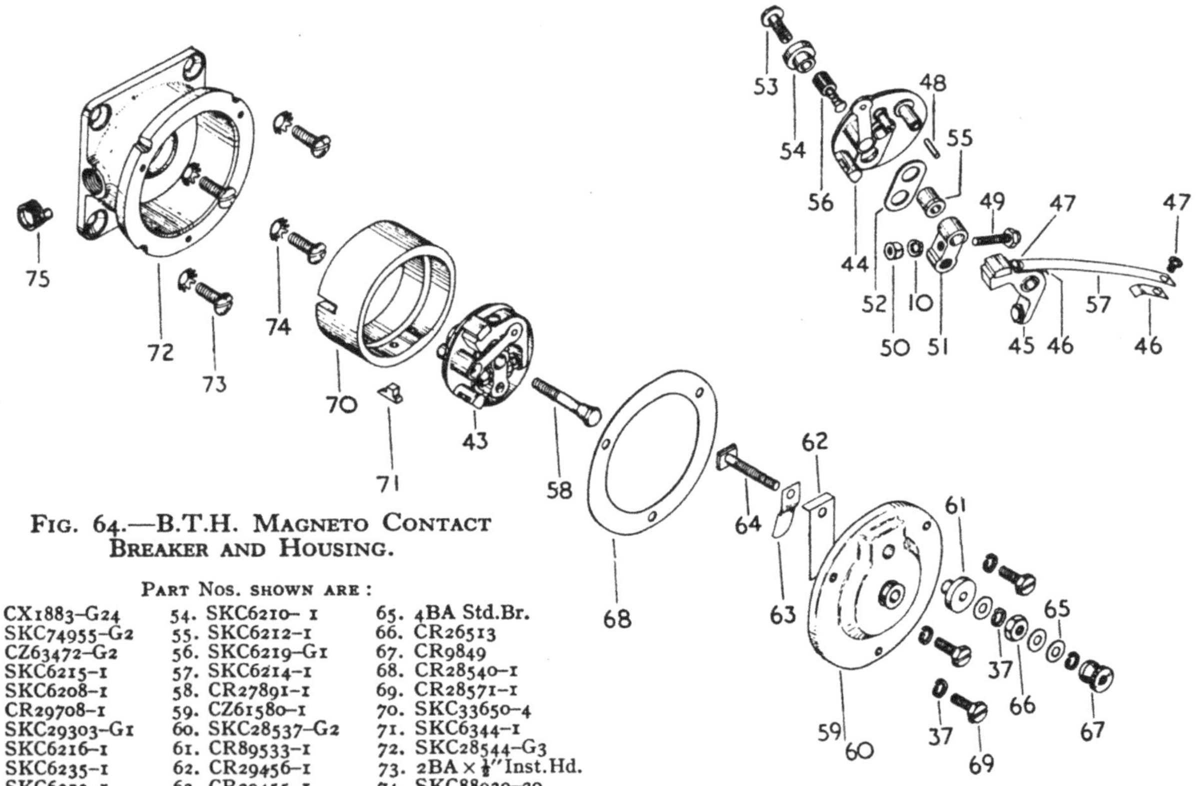

FIG. 64.—B.T.H. MAGNETO CONTACT BREAKER AND HOUSING.

PART NOS. SHOWN ARE:

43. CX1883–G24	54. SKC6210–1	65. 4BA Std.Br.
44. SKC74955–G2	55. SKC6212–1	66. CR26513
45. CZ63472–G2	56. SKC6219–G1	67. CR9849
46. SKC6215–1	57. SKC6214–1	68. CR28540–1
47. SKC6208–1	58. CR27891–1	69. CR28571–1
48. CR29708–1	59. CZ61580–1	70. SKC33650–4
49. SKC29303–G1	60. SKC28537–G2	71. SKC6344–1
50. SKC6216–1	61. CR89533–1	72. SKC28544–G3
51. SKC6235–1	62. CR29456–1	73. 2BA × ½″ Inst.Hd.
52. SKC6213–1	63. CR29455–1	74. SKC88939–20
53. SKC6209–1	64. CR29459–1	75. CR26381–1

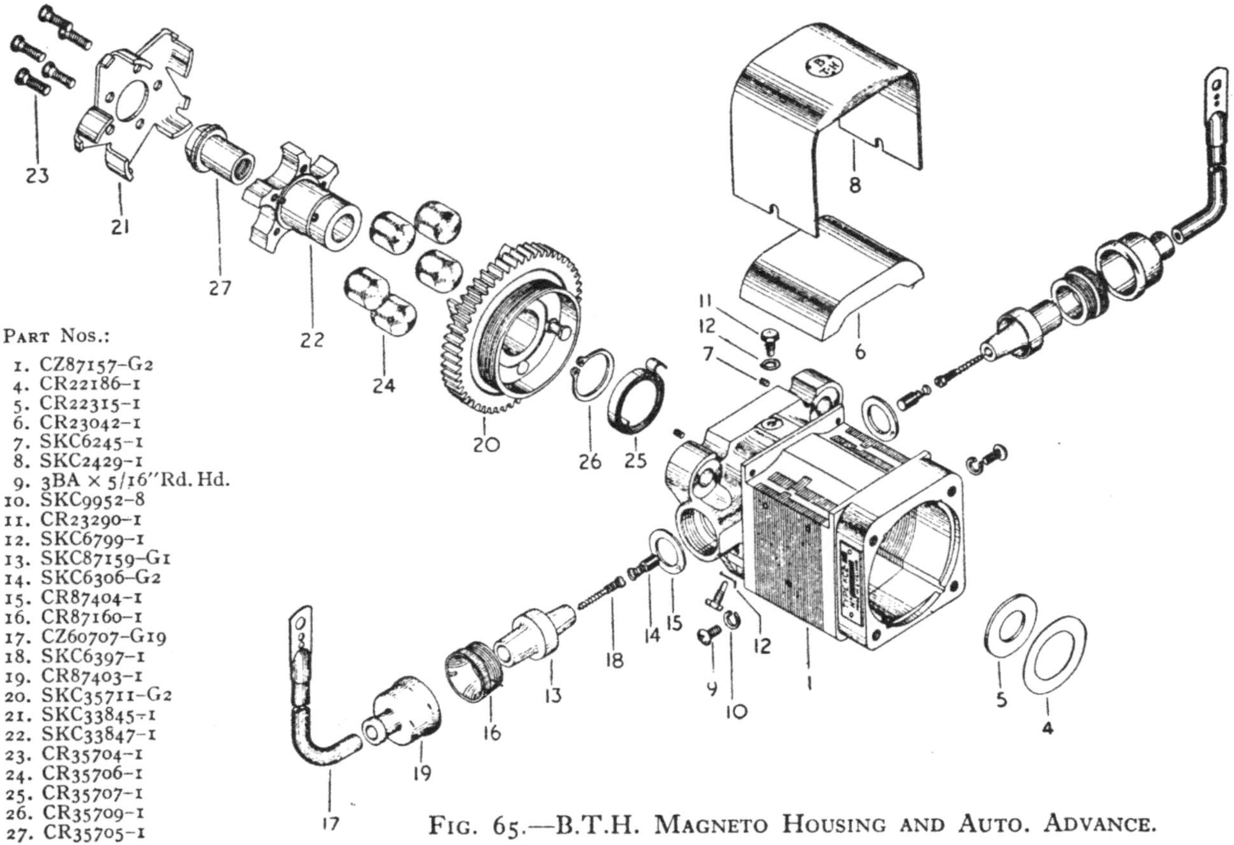

FIG. 65.—B.T.H. MAGNETO HOUSING AND AUTO. ADVANCE.

PART NOS.:

1. CZ87157–G2
4. CR22186–1
5. CR22315–1
6. CR23042–1
7. SKC6245–1
8. SKC2429–1
9. 3BA × 5/16″ Rd. Hd.
10. SKC9952–8
11. CR23290–1
12. SKC6799–1
13. SKC87159–G1
14. SKC6306–G2
15. CR87404–1
16. CR87160–1
17. CZ60707–G19
18. SKC6397–1
19. CR87403–1
20. SKC35711–G2
21. SKC33845–1
22. SKC33847–1
23. CR35704–1
24. CR35706–1
25. CR35707–1
26. CR35709–1
27. CR35705–1

Reassembly of Magneto

Reassembly of the magneto is exactly reverse to the foregoing instructions, but the following points must be observed:

(1) If the armature is withdrawn from the housing it will be necessary to remagnetise the magnet after replacing the armature.

(2) Make certain that all parts are thoroughly cleaned and that no foreign matter exists, particularly in the tunnel of the main housing, otherwise binding and ultimate seizure of the armature will follow.

(3) It is essential that all washers and locking tabs should be placed in the original position, and damaged tabs should not be used, but replacements obtained.

(4) If the cam ring has been removed from the contact end-plate, be sure that the lubricating wick is in position and lubricated with thin oil, and that the cam is properly located in the end-plate with the timing slot engaged in the timing top screw. In no circumstances should the timing top screw be removed from its original position, as this is set by an electrical timing process at the Works during manufacture.

(5) Care should be taken when replacing the contact breaker to ensure that the key on the contact-breaker base engages with the corresponding slot in the armature and held firmly in this position during the tightening of the retaining screw.

(6) All carbon brushes should move freely in their holders and the spring pressure should not be adjusted unnecessarily.

(7) High-tension cables should be examined for cracks, such as dry or perished rubber, and should be screwed securely home in the collector mouldings. A special screw is fitted in the collector mouldings for this purpose and has a wood, screw-threaded portion protruding in the cable outlet position into which the cables are screwed.

Magneto Lubrication

The magneto armature runs on ball bearings which are packed with grease before being put into service. It will, therefore, not be necessary to lubricate these bearings until the magneto is overhauled at approximately 20,000 miles.

Contact Breaker

At intervals, the magneto contact breaker should be inspected, as it is important that the points are kept clean and free from oil or grease. If the points are left in a dirty state, they will become oxidised, causing bad electrical contact between the points when closed and also result in difficult starting and misfiring. The points are made slightly convex and when necessary may be cleaned with a fine emery cloth or a carborundum stone—in no circumstances should they be filed. The contact breaker may be removed for cleaning by unscrewing the central hexagon-headed screw and withdrawing the assembly from its housing. The rocker-arm may then be lifted from its bearing bush by first raising and then moving to one side the check spring which is located in the end of the bearing bush. Care should be taken not to distort the rocker-arm control-spring. When replacing the rocker-arm smear the bearing bush lightly with thin lubricating oil; carefully wipe off any surplus oil.

Collector Brush

The collector brush mouldings should be periodically removed and the surface wiped clean with a cloth moistened with petrol. Before replacing the mouldings the slip-ring track should be cleaned. To do this insert a corner of clean cloth into the aperture from which the mouldings have been taken, so that it bears against the slip-ring track and the flange; at the same time turn the magneto. This will remove any oil or carbon deposit which would be likely to cause an electrical leakage. See that the cleaned parts are dry and that the petrol vapour is expelled before restoring the magneto to service.

Automatic Timing Device

This device is incorporated with the magneto driving gear and does not require any attention or adjustment. It automatically provides the required magneto timing in relation to the engine speed and so considerably improves the general performance of the motor cycle. When stationary, and at idling speeds, the auto advance device automatically returns to the retard position, advancing as the speed of the engine increases.

Magneto Removal

In order to remove the magneto, it is necessary to take off the timing-case cover. To remove the magneto driving gear and automatic advance and retard mechanism, place a $\frac{5}{16}$ in. B.S.F. nut or a piece of metal of similar depth between one of the fixed bob-weight lips and a movable lip. This will prevent the rotatable mechanism coming in contact with the fully advanced stop. If this precaution is not taken the fibre wheel may be damaged beyond repair. Unscrew the shaft nut with a slow but firm turning movement (right-hand thread) which will release it from the shaft and at the same time withdraw the assembly. The shaft has a keyway cut in it, but no key is fitted. Remove the two H.T. leads from the plug end and disconnect the cut-out lead. After removing the three clamping nuts the magneto can be taken off.

Magneto Installation

Fit the magneto to the crankcase over the three studs, assemble the three nuts and tighten up. Fit the auto advance and retard mechanism gear; time the magneto to engine and tighten up the central nut.

Refit contact-breaker cover and H.T. leads to sparking-plugs. Clean timing cover and crankcase faces and smear jointing compound lightly on timing cover face before replacing.

Ignition Timing

If a depth gauge of suitable size is not available, it is convenient to make a gauge from a piece of spoke, so that the piston position can be accurately gauged. The spoke should be about 5 in. long and must be graduated in sixteenths of an inch with a small fine file. The ignition on the 350-c.c. and 500-c.c. machines is correctly timed when, with the magneto fully advanced and the contact-breaker points just breaking, the off-side piston is :—

350 c.c.	$\frac{11}{32}$ in. B.T.D.C.
500 c.c.	$\frac{3}{8}$ in. B.T.D.C.

FIG. 66.—GRADUATED SPOKE USED FOR DETERMINING THE PISTON POSITION FOR IGNITION TIMING.

FIG. 67.—FIXING THE B.T.H. AUTOMATIC ADVANCE ASSEMBLY IN THE ADVANCE POSITION FOR TIMING THE IGNITION.

In order to time the magneto in the fully advanced position it is necessary to wedge the automatic control. This can be carried out by turning the mechanism in an anti-clockwise direction, against the spring, and inserting the tool-kit screwdriver between one of the bob-weights and retard control lip. With the screwdriver in position make sure that the mechanism is touching the fully advanced stop (see illustration). (When a Lucas magneto is fitted remove the outer cover and place a suitable wedge between the stops, in order to fully advance the magneto points.) Take off the contact-breaker cover and rotate the magneto in the direction of rotation (anti-clockwise as viewed from the drive end) until the contact points are just opening.

Remove the sparking-plug and turn the engine by engaging a gear and with the machine on the stand rotating the rear wheel in the correct direction for forward travel. When both offside valves are closed on the compression stroke, bring the offside piston up to T.D.C. and insert the gauge through the sparking-plug hole to test the position. First making certain the piston is on true T.D.C., rotate the engine backwards until the piston has dropped $\frac{11}{32}$ in. or $\frac{3}{8}$ in., as the case may be. Now check up to see that the magneto is correctly positioned with the points just breaking and carefully tighten up the nut on the armature shaft, until a grip is felt. Immediately release the screwdriver from the mechanism, otherwise serious damage will result. With the screwdriver removed tighten up the shaft nut. Finally, turn over the engine a few times and then make a check by again positioning the piston at the appropriate measurement before T.D.C., and ascertaining that the contact-breaker points are just opening in the fully advanced position.

CHAPTER VIII

LUCAS ELECTRICAL EQUIPMENT

It is proposed to deal first with the Lucas electrical equipment on the latest twin-cylinder models; these are fitted with separate dynamo and magneto. The final part of the chapter is devoted to the " Magdyno " equipment fitted to the majority of earlier Triumph motor cycles.

IGNITION

Lucas Magneto

The magneto is fitted with a timing control which automatically varies the firing point according to the speed of the engine, thus relieving the rider of the necessity for adjusting the timing. Its advantages are particularly evident when accelerating and during hill climbing, the danger of pre-ignition, knocking or pinking being very much reduced.

The control consists of a driving gear carrying a plate fitted with two pins. A weight is pivoted on each pin and the movement of the weight is controlled by a spring connected between the pivot end of the weight and a toggle lever pivoted at approximately the centre of the weight. Holes are provided in each toggle lever, which locate pegs on the underside of a driving plate secured to the magneto spindle. This plate is also provided with stops which limit the range of the control.

When the magneto is stationary the weights are in the closed position, and the magneto timing is retarded for starting purposes. As the speed is increased, centrifugal force acting on the weights overcomes the restraining influence of the

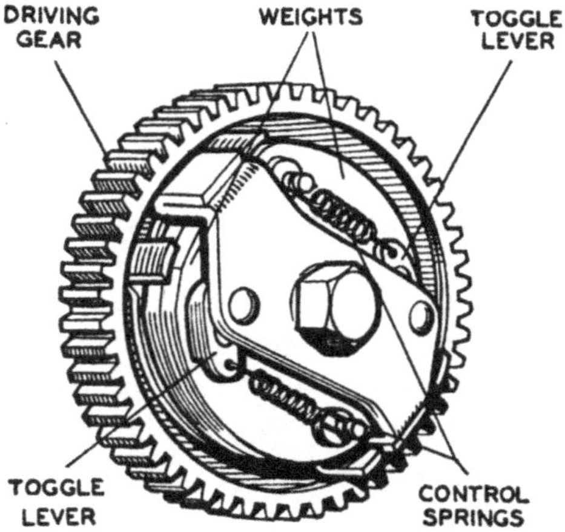

Fig. 68.—Automatic Advance Mechanism.

springs and the weights move outwards, causing relative movement to take place between the driving gear and the magneto spindle, so advancing the timing. By careful design of the control springs, the control is arranged to conform closely to the engine requirements.

Automatic advance begins at 500–600 r.p.m. (magneto) and gives maximum advance of 15 to 18 degrees at 1000–1100 (magneto) r.p.m.

Routine Maintenance every 3,000 miles—Lubrication

(1) *Cam.*—The cam is lubricated by a felt pad which is contained in a pocket in the cam housing. A small hole in the cam fitted with a wick enables the oil to find its way on to the surface of the cam. Remove the contact-breaker cover and turn the engine over until the hole on the face of the cam can be clearly seen and then carefully add a few drops of thin machine oil. Do not allow any oil to get on the contacts. When the magneto is dismantled for any reason the felt pad can then be removed and soaked in thin machine oil.

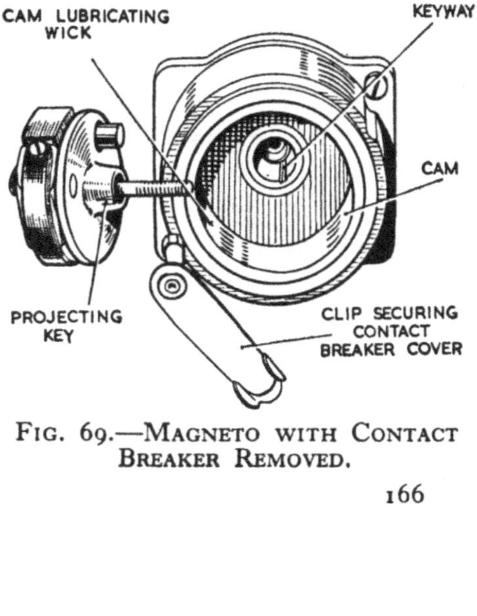

Fig. 69.—Magneto with Contact Breaker Removed.

(2) *Rocker Arm Pivot.*—To gain access to the rocker arm pivot, remove the complete contact breaker by taking out the

hexagon-headed screw from the centre of the contact breaker and pulling the contact breaker off the tapered shaft on which it fits. Then push aside the rocker-arm locating-spring, prise the rocker arm off its bearing and lightly smear the bearing with clean engine oil. At the same time lightly smear the contact breaker spring with clean engine oil.

When replacing the contact breaker, take care to ensure that the projecting key, on the tapered portion of the contact breaker base, engages with the keyway cut in the armature spindle, otherwise the timing of the magneto will be upset. Tighten the hexagon-headed screw with care; it must not be too slack, nor must undue force be used.

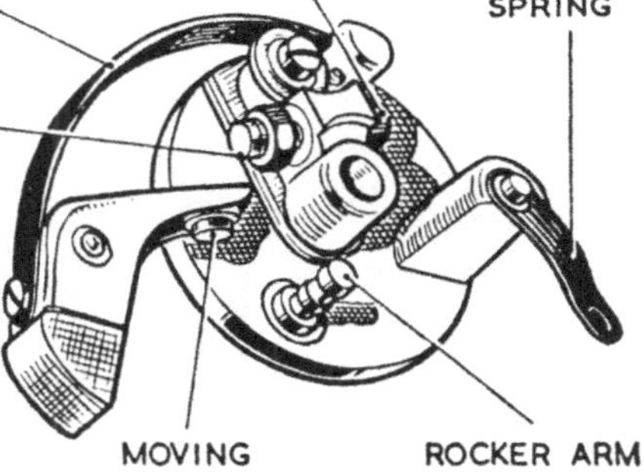

FIG. 70.—CONTACT BREAKER WITH MOVING CONTACT REMOVED.

Routine Maintenance—Adjustment

Remove the contact-breaker cover and turn the engine over until the contacts are fully opened. Check the gap with the gauge on the spanner provided in the tool-kit. The gauge has a thickness of 0·012 in. and if the setting is correct the gauge should be a sliding fit. If, however, the gap varies appreciably from the gauge, the contact breaker must be adjusted. To adjust, keep the engine in the position to give maximum opening of the contacts, slacken the lock-nut and turn the contact screw by its hexagon head until the gap is set to the gauge. Finally, tighten the lock-nut and recheck the setting.

Every 6,000 Miles

Contact Breaker Cleaning.—Remove the contact-breaker cover and examine the contact breaker. If the contacts are burned or blackened clean them with fine carborundum stone or fine emery cloth and afterwards wipe away any dust or dirt with a petrol-moistened cloth. Cleaning of the contacts is made easier if the contact breaker is removed as described above.

Testing in Position to Locate Cause of Misfiring or Failure of Ignition

(1) Disconnect the cable from one of the sparking-plugs and hold the end about ⅛ in. from some part of the cylinder block while the engine is turned over. If the sparking is strong and regular, the fault lies in the sparking-plug which must be removed for examination and if necessary cleaned and adjusted. Check the other sparking-plug in the same way.

(2) Next examine the high-tension cables. They may have become cracked or perished and may be sparking to the engine or frame. Correct by renewing; 7 mm. rubber-covered ignition cable must be used. To fit, bare the cable for about ¼ in. and thread it through the moulded terminal nut; pass the wire through the washer removed from the original cable and bend back the wire strands.

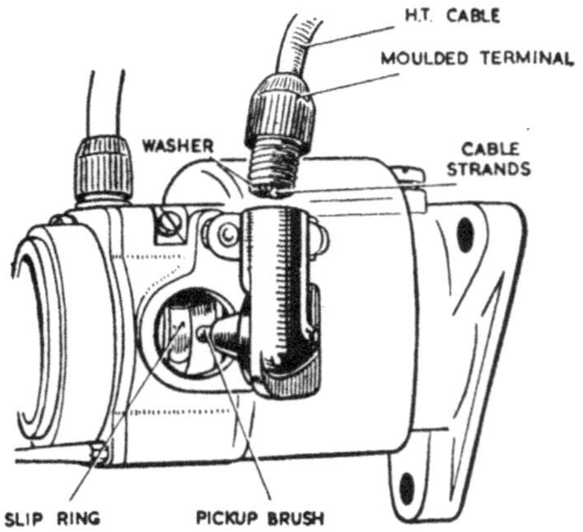

FIG. 71.—MAGNETO WITH PICK-UP AND HIGH-TENSION TERMINAL REMOVED.

If the performance of the magneto is still unsatisfactory, the contact breaker may require cleaning or adjustment (see above) or there may be an internal fault in the magneto. The following procedure should, therefore, be adopted.

(3) *Contact Breaker.*—Examine the contact breaker. If the contacts are badly burned they should be renewed by a replacement contact set. To remove the fixed contact slacken the lock-nut and remove the contact by turning the hexagon head. The new moving contact will be supplied complete with a new rocker-arm. When new contacts are fitted, or sometimes after cleaning, the gap between the contacts will need to be reset. Examine the contact-breaker spring and wipe away any dirt or rust.

(4) *High tension pick-ups and slip ring.*—Push aside the securing clips and remove the high-tension pick-ups. Wipe

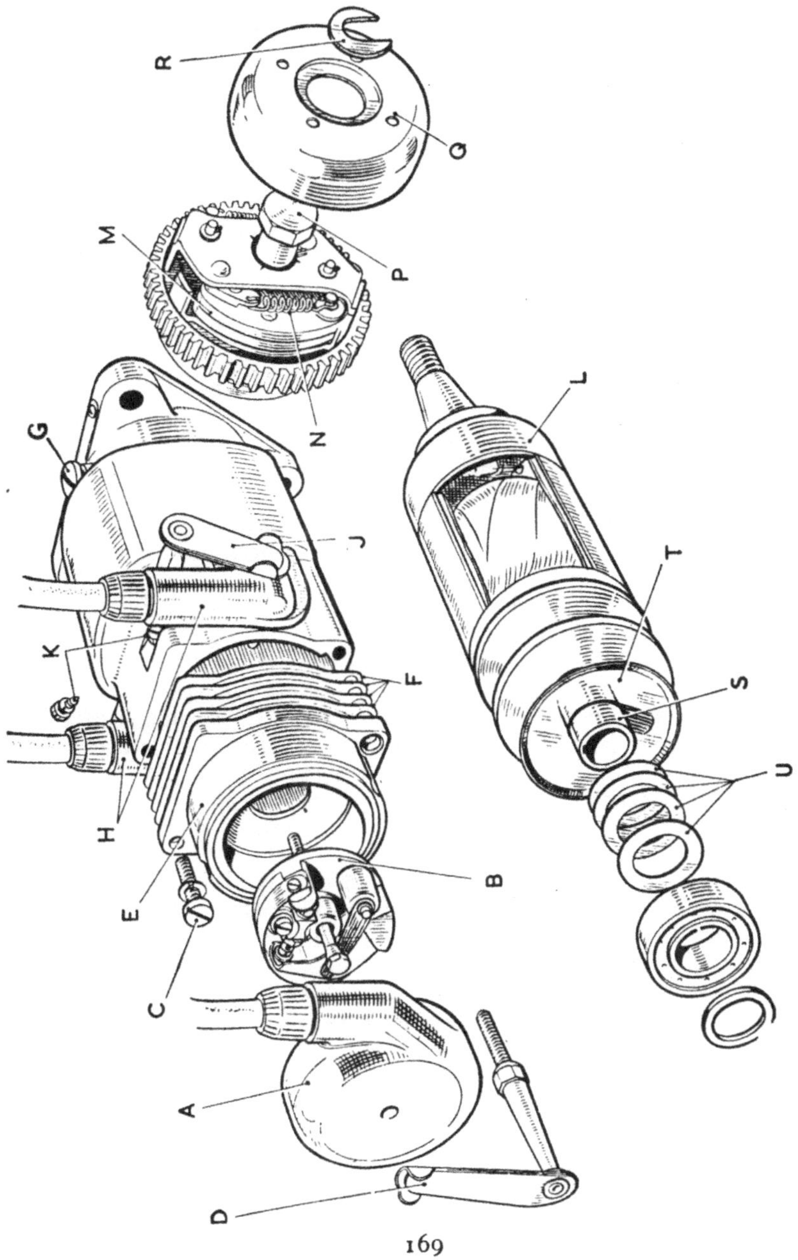

FIG. 72.—MAGNETO DISMANTLED.

them clean and polish with a fine dry cloth. The high-tension pick-up brush must move freely in its holder. If it is dirty, clean with a cloth moistened with petrol. If either of the brushes is worn to within ⅛ in. of the shoulder it must be renewed. While the high-tension pick-ups are removed, clean the slip-ring track and flanges by holding a soft cloth on the ring by means of a suitably shaped piece of wood, while the engine is slowly turned by hand.

Removal of magneto from engine.—See page 162.

Dismantling Magneto (see Fig. 72)

(1) Remove contact-breaker cover (*A*) by pushing aside securing clip (*D*).

(2) Remove complete contact-breaker assembly (*B*) by taking out hexagon-headed screw and pulling the contact breaker off the tapered shaft on which it fits. Remove high-tension pick-up assemblies (*H*) by pushing aside the spring securing clips (*J*). Remove screw (G) securing brush spring and holder.

(3) Remove the two hexagon-headed screws (*K*) (these are safety-gap screws and are positioned near the high-tension pick-up terminals).

(4) Remove the contact-breaker housing (*E*) from the magneto body by withdrawing the securing screw (*C*) and spring pillar (*D*) (take care not to lose the shims (*F*) which fit between the housing and the magneto body). Remove the automatic advance mechanism (*M*) from the magneto by slackening the securing bolt (*P*). The movement of this bolt forces the automatic advance mechanism off the armature shaft.

(5) The armature (*L*) complete with bearings, condenser and slip-ring can be removed from the magneto body by pushing on the armature shaft extension.

(6) The ball cages can be levered off the inner journals by means of a screwdriver. The inner journals can be removed from the armature by means of an extractor; care should be taken when removing the journal from the contact breaker end to see that the extractor bears on the brass extension (*S*) and not on the electric contact or insulator down the centre of the brass extension. It is suggested that a disc of appropriate diameter be placed across the face of the shaft extension.

When replacing, take care to refit grease-flinging plate (*T*) and also the full number of shims (*U*). A tool of the type shown should be used to remove the outer journals.

(7) *Automatic Timing Control.*—If it becomes necessary at any time to renew the automatic advance mechanism springs

(*N*), the cover (*Q*) of the advance mechanism can be pulled off when the washer (*R*) is removed from underneath the head of the bolt. Make sure that the springs are the correct type, otherwise the automatic advance will upset the engine performance.

Reassembly of Magneto

Wash the bearings in white spirit, dry thoroughly and repack with high-melting-point grease. Fit ball races on armature shaft by means of a hand-press and use a mandrel of the type illustrated to fit the outer bearings. It will be necessary to renew the rubber seal before replacing outer bearing at driving end.

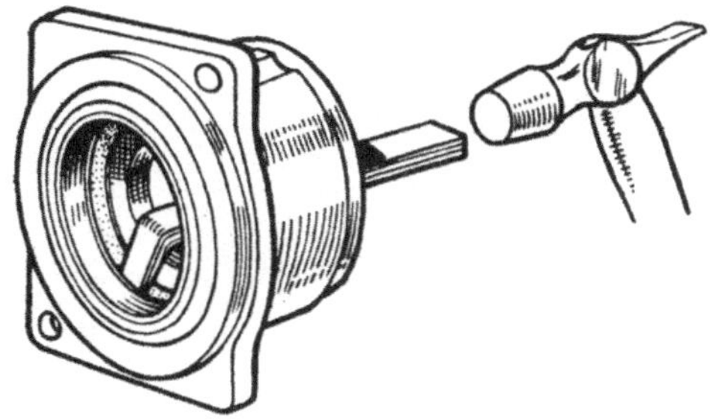

FIG. 73.—TOOL FOR REMOVAL OF OUTER MAGNETO BALL BEARING JOURNALS.

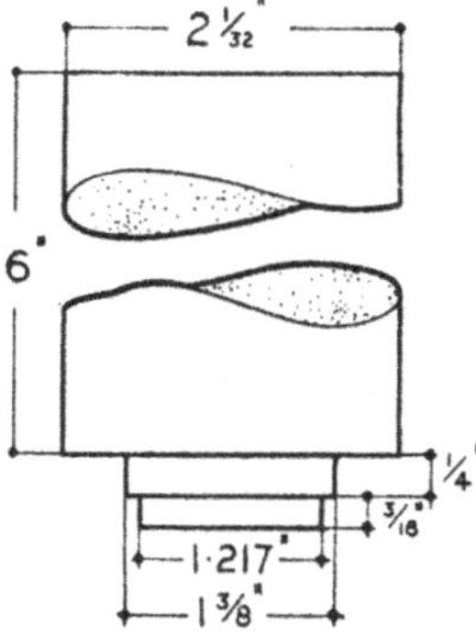

FIG. 74.—MANDREL FOR FITTING OUTER BALL BEARING JOURNALS.

Refit the armature in the magneto body. Now refit the contact-breaker housing, taking care to replace all the shims. (*Note.*—If a replacement armature is being fitted, it may be necessary to alter the number of shims.) Before replacing any other parts make sure that the armature revolves freely. Now refit contact-breaker assembly after lubricating the contact-breaker spring and rocker-arm pivot, taking care that the projecting key on the tapered shaft fits into the keyway cut in the armature shaft. Tighten the hexagon-headed screw with care. It must not be too slack, nor must undue force be used.

Refit earthing screw complete with brush and spring. Refit hexagon-headed safety-gap screws and high-tension pick-up assemblies.

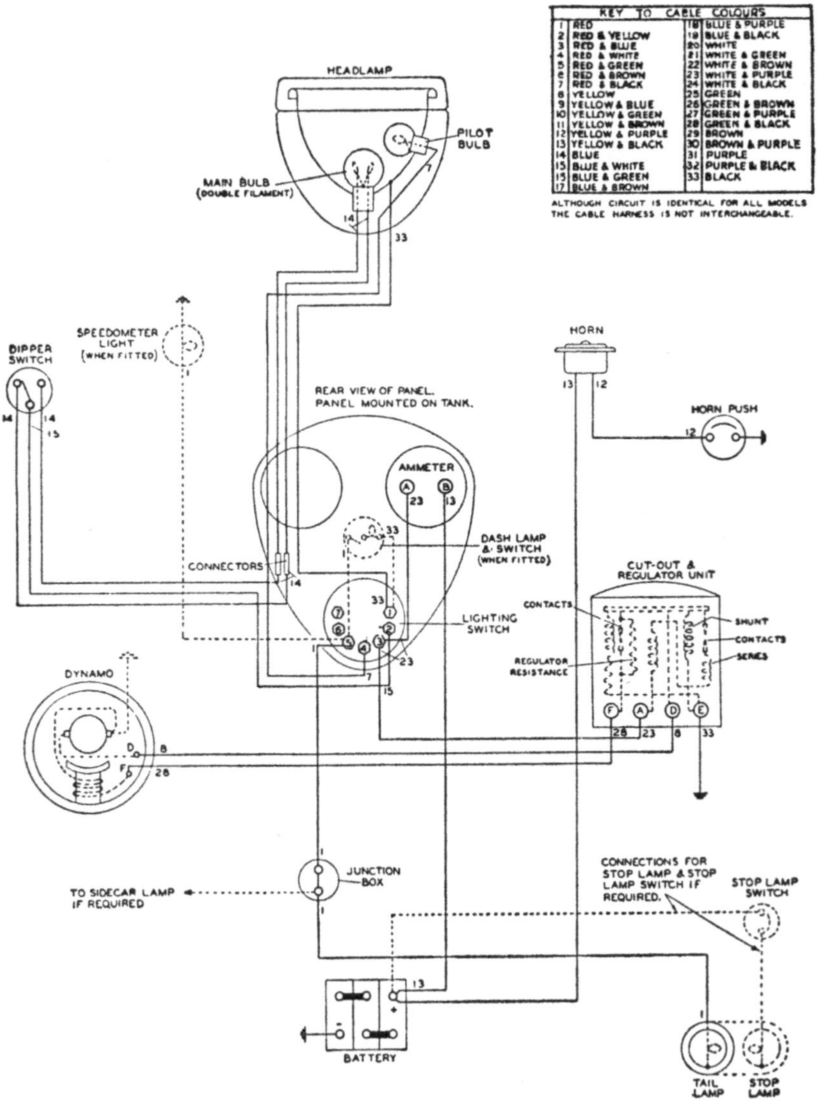

Fig. 75.—Wiring Diagram of Lucas Electrical Equipment with Separate Dynamo and Voltage Control Regulator.

Refitting Magneto and Ignition Timing

The instructions for refitting the magneto and ignition timing are as described on pages 162 and 163.

CHARGING

Dynamo and Output Control

The dynamo is of the compensated voltage-control type and works in conjunction with a regulator unit which is mounted together with the cut-out. The regulator and cut-out units are accurately set and should not be tampered with.

The regulator provides a completely automatic control. It causes the dynamo to give an output which varies according to the load on the battery and its state of charge. When the battery is discharged the dynamo gives a high output, but if the battery is fully charged, then the dynamo gives only a trickle charge so as to keep the battery in a good condition without danger of overcharging. In addition to controlling the output of the dynamo according to the condition of the battery, the regulator provides for an increase of output to balance the current taken by lamps when in use.

Dynamo Test Data

Cutting-in speed 1250–1500 r.p.m. at 7 dynamo volts. Output test 5 amps. at 2000–2500 r.p.m. at 7 dynamo volts taken on 1·4 ohm resistance load without regulator.

Routine Maintenance

Lubrication.—The lubricator at the commutator end bracket must be given a few drops of good grade thin machine oil every 4,000–5,000 miles. The bearing at the driving end is packed with grease and will last until the machine is taken down for a general overhaul.

Inspection of Commutator and Brushgear.—About once every six months remove the dynamo cover band for inspection of commutator and brushes.

The brushes must make firm contact with the commutator. The brushes are held in boxes by means of springs; move each brush to see that it is free to slide in its holder; if it sticks, remove it and clean with a cloth moistened with petrol. Care must be taken to replace the brushes in their original positions, otherwise they will not "bed" properly on the commutator. If after long service the brushes have become worn to such an extent that the brush flexible is exposed on the running face or if the brushes do not make good contact with the commutator, they must be replaced by genuine Lucas brushes. The commutator should be free from any trace of oil or dirt and should have a highly polished appearance. Clean a dirty or blackened commutator by pressing against it a fine dry cloth while the engine is slowly turned over by hand. If the commutator is very dirty, moisten the cloth with petrol.

Testing the Dynamo in Position

(1) Check that the dynamo and regulator unit are connected correctly. The dynamo terminal (*D*) should be connected to the regulator unit terminal (*D*), and dynamo terminal (*F*) to regulator unit terminal (*F*).

(2) Remove the cables from the dynamo terminals (*D*) and (*F*) and connect the two terminals with a short length of wire.

(3) Start the engine and set to run at normal idling speed.

(4) Connect the positive lead of a moving coil voltmeter, calibrated 0–10 volts, to one of the dynamo terminals and connect the negative lead to a good earthing point on the dynamo yoke or engine.

(5) Gradually increase the engine speed, when the voltmeter reading should rise rapidly and without fluctuation. Do not allow the voltmeter reading to rise above 10 volts. Do not race the engine in an attempt to increase the voltage. It is sufficient to run the dynamo up to a speed of 1000 r.p.m. If there is no reading, check the brush gear. If there is a low reading of approximately ½ volt, the field winding may be at fault. If there is a reading of approximately 1½ to 2 volts, the armature winding may be at fault.

(6) Remove the dynamo cover band and examine the brushes and commutator. Hold back each of the brush springs and move the brush by pulling gently on its flexible connector. If the movement is sluggish, remove the brush from its holder and ease the sides by lightly polishing on a smooth file. Always replace brushes in their original positions. If the brushes are

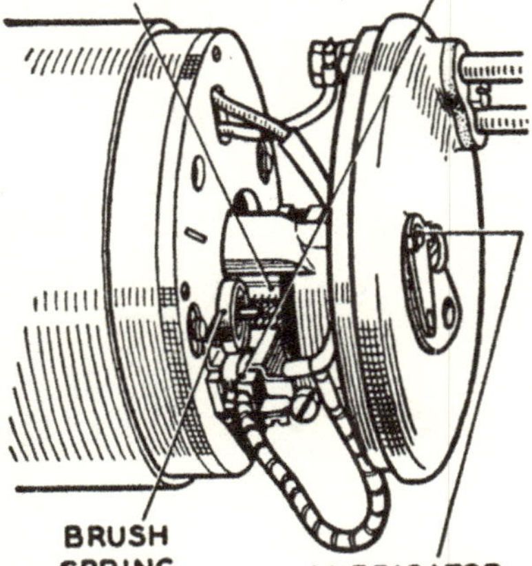

FIG. 76.—DYNAMO COVER BAND REMOVED.

worn so that they do not bear on the commutator or if the brush flexible is exposed on the running face, new brushes must be fitted.

If the commutator is blackened or dirty, clean it by holding a petrol-moistened cloth against it while the engine is turned slowly by hand.

Re-test the dynamo ; if there is still no reading on the voltmeter, there is an internal fault and the complete unit, if a spare is available, should be replaced.

(7) If the dynamo is in good order, restore the original connections to the dynamo. Connect regulator unit terminal (*D*) to dynamo terminal (*D*) and regulator terminal (*F*) to dynamo terminal (*F*). Remove the lead from the (*D*) terminal on the regulator unit and connect the voltmeter between this cable and an earthing point on the engine. Run the engine as before. The reading should be the same as that measured directly at the dynamo. No reading indicates a break in the cable to the dynamo. If the reading is correct, test the regulator unit (see page 181).

Dynamo Dismantling (see Fig. 77)

(1) Bend back the tag on the washer (*B*) locking the screw (*A*) securing the driving gear (*C*) and remove the screw. Withdraw the gear from the dynamo shaft by carefully levering it off or by means of an extractor. Remove the keys (*D*) from the shaft.

(2) Remove the cover band (*H*), hold back the brush springs and lift the brushes from their holders.

(3) Take out the screw (*J*) with spring washer, from the centre of the black moulded end-cap (*G*). Draw the cap away from the end bracket, take off terminal nuts (*F*) and spring washers, and lift the connections off the terminals.

(4) Unscrew and remove from the driving end bracket the two through-bolts (*L*) securing the driving end bracket (*N*) and commutator end bracket (*Q*) to the dynamo yoke (*M*).

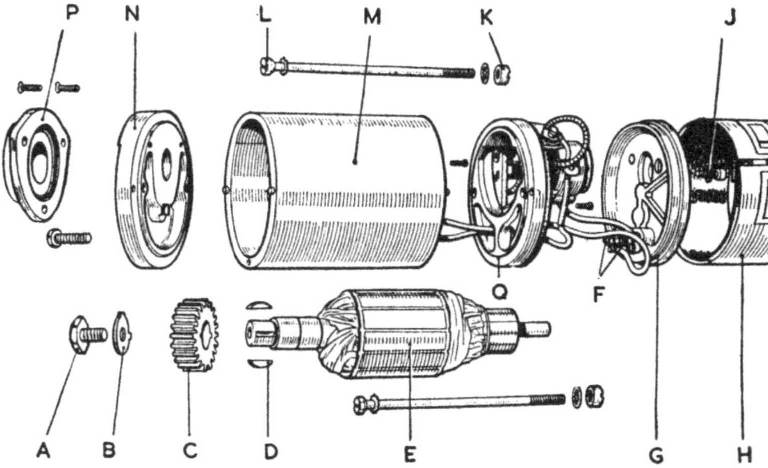

Fig. 77.—Dynamo Dismantled.

Hold the nuts (*K*) at the commutator end while unscrewing the bolts, and take care not to lose the nuts.

(5) Draw the driving end bracket complete with armature (*E*) out of the yoke.

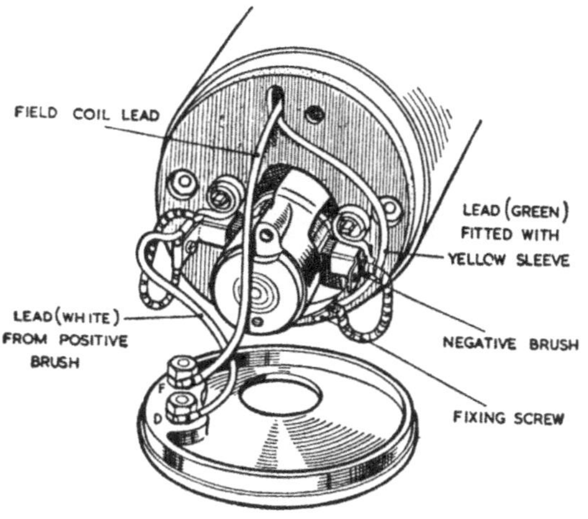

Fig. 78.—Dynamo Brush Connections.

Fig. 79.—Testing Brush Spring Tension.

(6) Remove bearing retaining plate (*P*) from driving end bracket which is secured by three screws (take great care with this plate as it contains an oil seal which must not be damaged). The armature and the oil seal collar can be withdrawn from the driving end bracket by means of a hand-press.

(7) Take out the screw securing the green field coil lead with the yellow sleeve to commutator end bracket and remove the end bracket, withdrawing the connectors through the slot in the insulating plate.

(8) Unscrew the three screws securing the insulating plate to the commutator end bracket and remove the plate complete with brush gear.

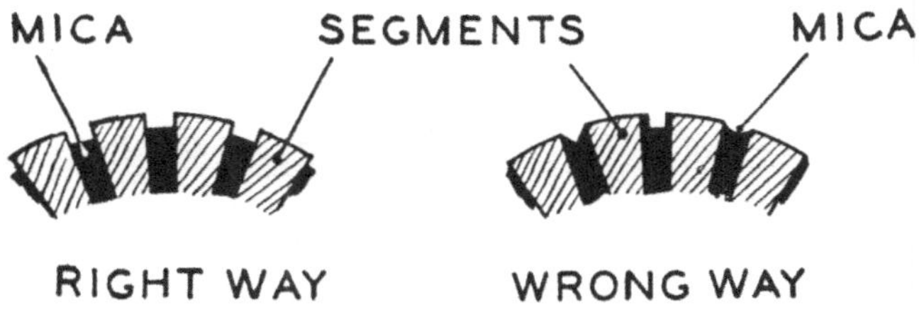

FIG. 80.—SHOWING RIGHT AND WRONG WAYS OF UNDERCUTTING MICA.

Dynamo Overhaul

(1) *Brushes.*—Test if brushes are sticking. Clean with petrol and if necessary ease the sides by lightly polishing on a smooth file. Replace brushes in their original positions.

Test the brush springs with a spring scale if available. The correct tension is 10–15 oz. Fit a new spring if the tension is low. If the brushes are worn so that the flexible is exposed on the running face, new brushes must be fitted. Brushes are pre-formed so that bedding to the commutator is unnecessary.

(2) *Commutator.*—A commutator in good condition will be smooth and free from pits or burned spots. Clean the commutator with a petrol-moistened cloth. If this is ineffective, carefully polish with a strip of very fine glass-paper while rotating the armature. To remedy a badly worn commutator, mount the armature with or without the drive end bracket in a lathe, rotate at high speed and take a light cut with a very sharp tool. Do not remove more metal than is necessary. Polish the commutator with very fine glass-paper.

Undercut the mica insulation between the segments to a depth of $\frac{1}{32}$ in. with a hacksaw blade ground down until it is only slightly thicker than the mica.

(3) *Field Coil.*—Test the field coil by connecting it in series with a 6-volt battery and a 6-volt, 3-watt bulb. If the field coil is satisfactory, the bulb should light up, but its brilliance should be somewhat less than when connected direct to the battery. Failure of the bulb to light indicates an open circuit in the field winding, while if the bulb lights up with full brilliance, the field coil is probably either shorted or earthed

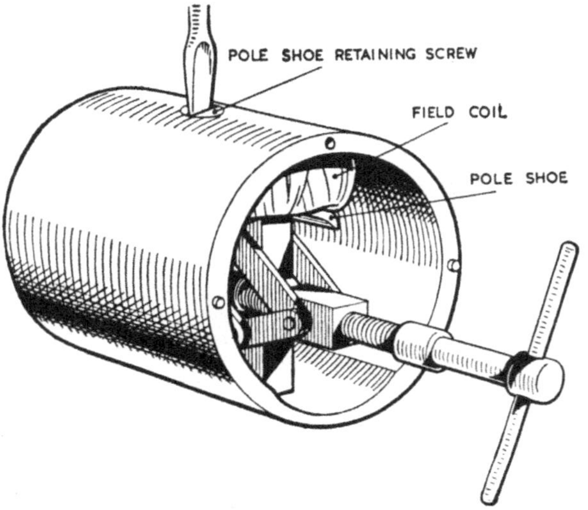

FIG. 81.—POLE SHOE AND FIELD COIL ASSEMBLY OF DYNAMO.

to the pole shoe or dynamo yoke. In either case, the complete dynamo assembly must be returned to a depot and a replacement fitted. If, however, a pole-shoe expander and a wheel-operated screwdriver are available, it is possible to replace the field coil. A pole-shoe expander is necessary to ensure that there will not be any air-gap between the pole shoe and the inner face of the yoke.

Replace the field coil as follows :—

Unscrew the pole-shoe retaining-screw (Fig. 81) by means of the wheel-operated screwdriver. Draw the pole-shoe and field coil out of the yoke and lift off the coil.

Fit the new field coil over the pole-shoe and place it in position inside the yoke. Take care to ensure that the taping of the field coil is not trapped between the pole shoe and the yoke.

Locate the pole shoe and field coil by lightly tightening the fixing screw. Insert the pole-shoe expander open to its fullest extent and tighten the screw. Remove the expander and give the screw a final tightening with the wheel-operated screwdriver. Lock the screw in position by caulking—that is, tapping some of the metal of the yoke into the slot in the head of the screw.

Armature.—The testing of the armature winding requires the use of a volt-drop test or a "growler." If these are not available, the armature should be checked by substitution. No attempt should be made to machine the armature core or to true a distorted armature shaft.

Dynamo Bearings.—A ball bearing is fitted at the driving end and a plain porous bronze bearing bush at the commutator end.

Bearings which are worn to such an extent that they will allow side movement of the armature shaft must be replaced.

To replace the bearing bush at the commutator end, remove the screw, press the bearing bush out of the commutator end bracket and remove the felt ring. Press the new bearing bush into the end bracket using a shouldered mandrel of the same diameter as the shaft which is to fit in the bearing. *Note.*—Before fitting the new bearing bush it should be allowed to stand for twenty-four hours immersed for about seven-eighths of its length in thin engine oil. The bush should be pressed in until it is flush with the face of the end bracket. Fit the felt ring in the space between the bearing and the wall of the bearing housing.

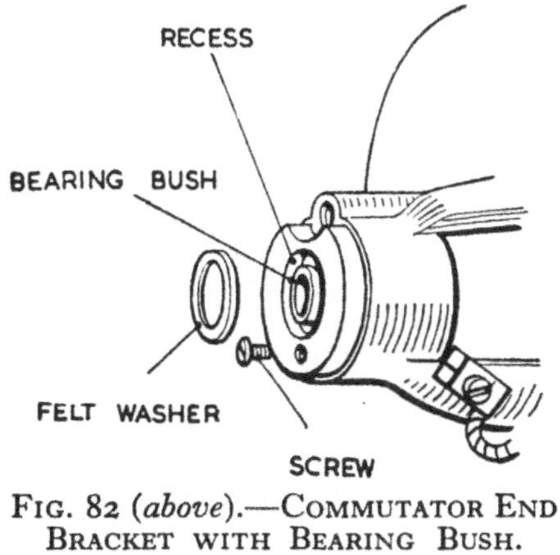

FIG. 82 (*above*).—COMMUTATOR END BRACKET WITH BEARING BUSH.

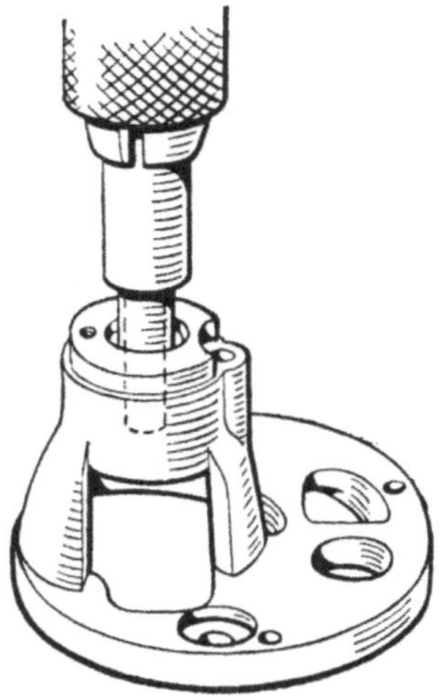

FIG. 83 (*right*).—FITTING BEARING BUSH USING SHOULDERED MANDREL.

The ball bearing at the driving end is replaced as follows: Remove bearing retaining plate from driving end bracket as previously described. Press the bearing out of the end bracket using a metal drift locating on the inner journal of the bearing. Wipe out the bearing housing and pack the new bearing with H.M.P. grease. Position the bearing in its housing and press it squarely home, applying pressure on the outer journal of the bearing.

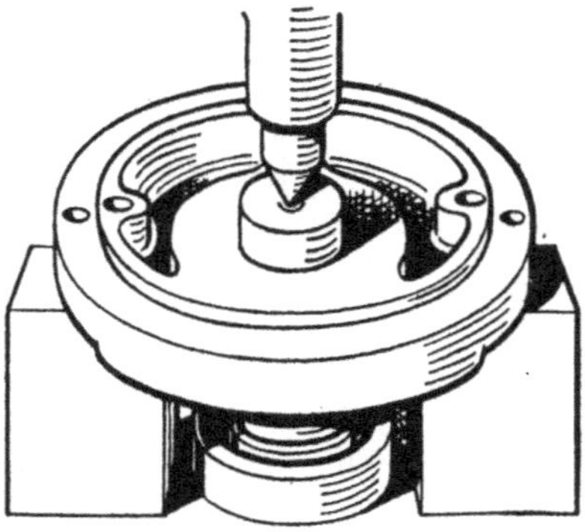

FIG. 84.—REMOVAL OF DYNAMO DRIVING-END BEARING.

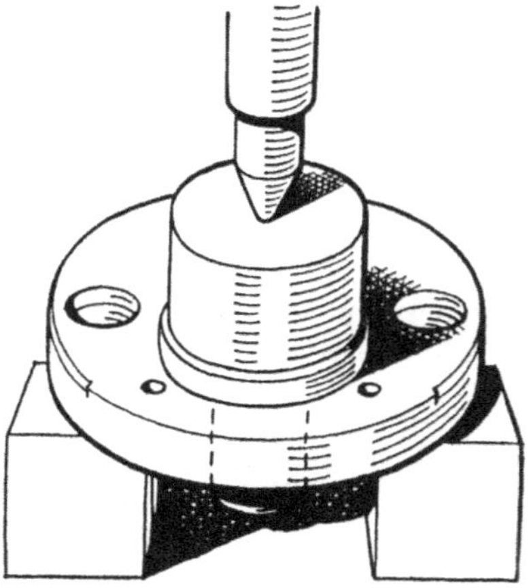

FIG. 85.—REPLACEMENT OF DRIVING-END BEARING.

Reassembly of Dynamo

In the main, the reassembly of the dynamo is a reversal of the operations described above, bearing in mind the following points:—

(1) The field coil fitted with the short length of yellow tubing, together with eyelet of the negative brush, must be secured to the commutator end bracket by means of the screw provided.

(2) The second field coil lead must be connected to terminal (*F*) on the moulded end-cap.

(3) The lead (coloured white) from the terminal on the positive brush box must be connected to terminal (*D*) on the moulded end-cap.

(4) Take care to refit dynamo cover band in original position and make sure that the securing screw does not short on dynamo brush gear.

Regulator Unit

This unit houses the dynamo voltage regulator unit and the cut-out. Both units are accurately set and the cover should be removed for cleaning and adjustment only in the event of trouble with the charging circuit being experienced.

Testing Regulator Unit in Position

(1) Before checking the regulator unit, make sure that the wiring between the regulator and the battery is in order. To do this, disconnect the wire from the (*A*) terminal of the regulator unit and connect the lead from the positive terminal of the voltmeter to the end of the wire. If a voltmeter reading is given, the wiring is in order and the regulator should be examined. If there is no reading examine the wiring for broken cables or loose connections.

(2) Remove the cable from the terminal on the regulator marked (*A*). Connect the positive terminal of the moving coil voltmeter to the (*D*) terminal on the regulator and connect the other lead of the voltmeter to an earthing point on the engine.

(3) Start the engine and slowly increase the speed until the voltmeter needle " flicks " and then steadies ; this should occur at a voltmeter reading between the limits given for the particular atmospheric temperature :—

Atmospheric Temperature.	Regulator Setting.
30° F.	7·9–8·3 volts
60° F.	7·8–8·2 volts
90° F.	7·7–8·1 volts

If the voltage at which the reading becomes steady is outside these limits, the regulator must be adjusted. Shut off the engine, release the lock-nut on the regulator adjusting screw and turn the screw in a clockwise direction to raise the setting, or in an anti-clockwise direction to lower the setting. Turn the screw a fraction of a turn at a time and then tighten the lock-nut.

When adjusting, do not run the engine up to more than half-throttle, as while the dynamo is on open circuit, it will build up to a high voltage if run at a high speed and so a false voltmeter reading would be obtained.

Cleaning Regulator Contacts

After long periods of service it may be found necessary to clean the vibrating contacts of the regulator. These are

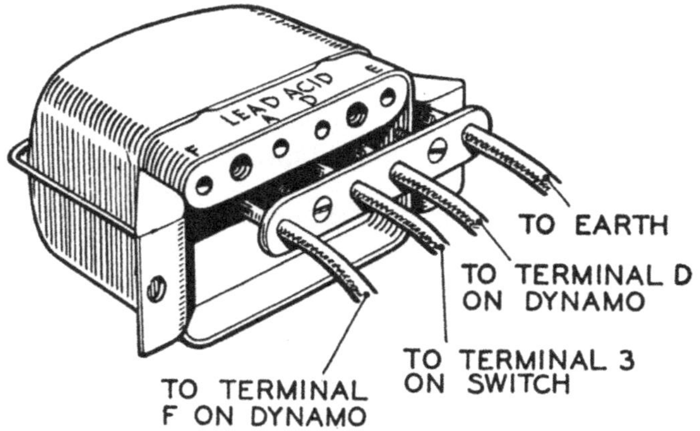

FIG. 86.—REGULATOR UNIT.

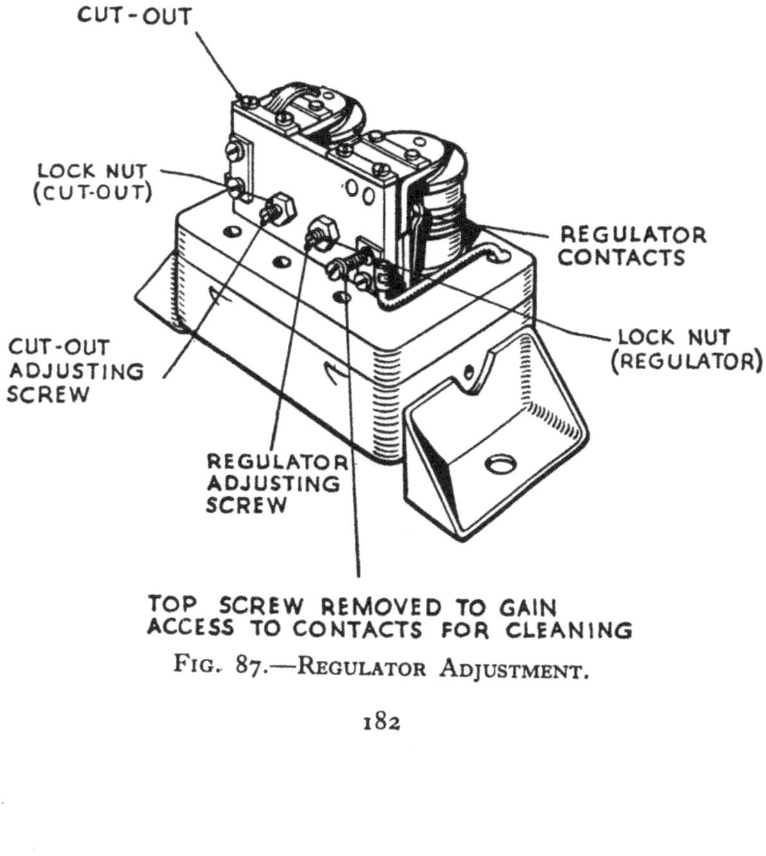

FIG. 87.—REGULATOR ADJUSTMENT.

accessible if the top screw securing the fixed contact is turned back and the bottom screw slackened to permit the fixed contact to be swung outwards. The contacts can then be polished with fine emery cloth.

Cut-Out—Testing and Adjustment

(1) If the regulator setting is within the correct limits, but the battery is still not receiving current from the dynamo, the cut-out may be out of adjustment, or there may be an open circuit in the wiring of the voltage regulator unit.

(2) Remove the voltmeter lead from the (*D*) terminal of the regulator unit and connect it to terminal (*A*). Run the engine as before: the reading on the voltmeter should be the same as that obtained when the voltmeter was connected to terminal (*D*). If there is no reading, the setting of the cut-out may be badly out of adjustment and the contacts are not closing.

(3) To check the voltage at which the cut-out operates, the voltmeter should be connected between the (*D*) terminal and earth. Slacken the lock-nut on the cut-out adjustment screw and turn the screw in an anti-clockwise direction until the cut-out contacts are seen to close. Check the voltage at which the cut-out operates and if necessary adjust by turning the screw in a clockwise direction to raise the setting or anti-clockwise direction to lower it. Set the cut-out so that it operates at 6·2–6·6 volts. Tighten the lock-nut after making the adjustment.

Cut-Out—Contacts Burnt or Dirty

To clean the cut-out contacts, place a strip of fine glass-paper between the contacts, then, with the contacts closed by hand, draw the paper through. This should be done two or three times with the rough side towards each contact.

Ammeter—Testing in Position

To test the ammeter in position, take out the screw securing the panel on the tank, disconnect oil pipe and remove the panel. With the engine stationary, check the voltage between each terminal of the ammeter and earth. Both readings should be the same. If there is a reading at terminal (*B*), but not at terminal (*A*), there is a broken connection in the ammeter and a replacement must be fitted.

To remove the ammeter, unscrew the ammeter terminal screws and lift out the cable ends. Bend back the four metal tags securing the ammeter and remove it from the panel. The procedure must be reversed when fitting the replacement ammeter.

Battery Maintenance

A 6-volt 12-ampere-hour lead–acid battery is fitted. The negative terminal is earthed to the frame.

For special precautions necessary at temperatures below 32° F. (freezing point), see later. When atmospheric temperature exceeds 100° F., examine battery at intervals not exceeding seven days, and at temperatures below this, every fourteen days, as follows :—

(1) Remove the battery cover, brush dirt from top of battery and remove vent plugs. *Never* bring a naked light near a battery when the vent plugs have been removed, or when the battery is being charged, as the gas given off by the electrolyte is highly explosive.

(2) *The Specific Gravity of Electrolyte* indicates the state of charge of the battery. Table " A " shows what the specific gravity of the electrolyte should be, at various temperatures, when the battery is fully charged. Check the gravity by means of a hydrometer, and if it is below the figures shown for the appropriate temperature in Table " B ", the battery should be charged as soon as possible by the normal running of the motor cycle. If this cannot be arranged, the battery should be charged from an external source.

If the level of the electrolyte is so low that a hydrometer reading cannot be taken, no attempt should be made to take a reading after adding distilled water until the battery has been on charge for at least thirty minutes.

Never transfer the electrolyte from one cell to another.

Never leave a battery in a discharged condition. It must be put on charge as soon as possible.

(3) *Electrolyte Level.* Check if the electrolyte in each cell is level with the top of the separators. Top up, if necessary, with distilled water. Do not allow distilled water to come in contact with metals—use only a glass or earthenware container and funnel.

If a battery is found to need an excessive amount of topping up, steps should be taken to find out the reason. For example, the battery may be receiving an excessive charge, in which case the regulator setting should be checked (see page 181). If one cell in particular needs topping up more than another, it is likely

SPECIFIC GRAVITY OF ELECTROLYTE AT VARIOUS TEMPERATURES.

TABLE "A".
(*Battery Fully Charged*)

1·270 at	120° F.
1·280 at	100° F.
1·285 at	80° F.
1·295 at	60° F.
1·305 at	40° F.
1·310 at	20° F.
1·320 at	0° F.
1·325 at	–21° F.

TABLE "B".
(*Battery Discharged*)

1·220 at	120° F.
1·230 at	100° F.
1·235 at	80° F.
1·245 at	60° F.
1·255 at	40° F.
1·260 at	20° F.
1·270 at	0° F.
1·275 at	–21° F.

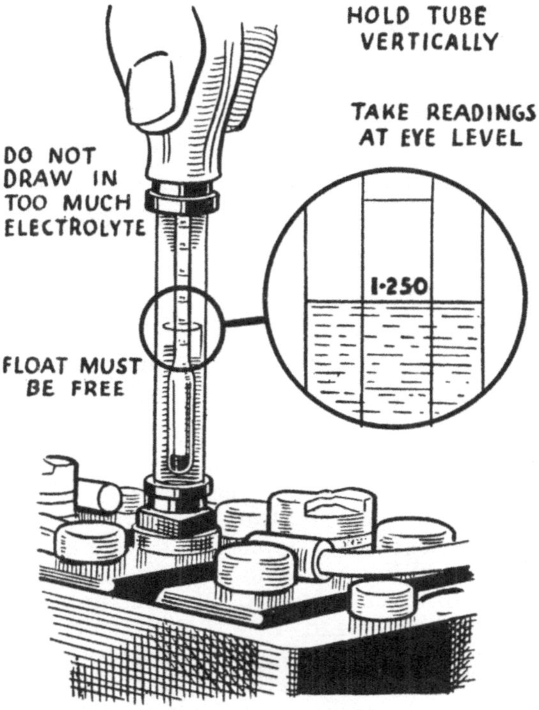

FIG. 88.—TAKING HYDROMETER READINGS.

that the case or container is cracked, in which event the battery must be replaced and arrangements made to clean up the battery carrier. Metal parts should be well cleaned and if possible washed with a solution of ammonia or bicarbonate of soda in water. After cleaning, paint the battery carrier and other surfaces affected by the electrolyte with anti-sulphuric paint, if available.

(4) Vent plugs should be kept clean and air passages free. Refix tightly after replacing the rubber washers.

(5) *Cleanliness.* Keep the battery and surrounding parts, particularly the tops of the cells, clean and dry. Brush away any sand, dust or road slush.

Battery electrolyte, which contains sulphuric acid in a diluted form, is destructive to practically everything except rubber, lead, glass or earthenware, therefore rags used to clean battery tops, etc., should be thrown away afterwards. If they are put in the tool box, for example, they will cause the tools to rust.

(6) *Fixing.* The battery must be held firmly in its mounting. Fixing bolts must be kept tight without using undue force.

Charging Battery from External Source

Before starting the charge, the battery should be topped up with distilled water to the top of the separators. The charge should be at the rate of 1·5 amps. and continued until the voltage and specific gravity of each cell show no further increase over four successive hourly readings, and all cells gas freely. If the specific gravity of the electrolyte in any cell or cells fails to rise whilst on charge and gassing does not take place, these should be tested for internal short circuit.

Temperature of Electrolyte During Charge.—The maximum permissible temperature of electrolyte during external charging is 120° F., and if this is reached, the charge should be suspended to allow the temperature to fall by at least 10° F. If charging from an external source has to be carried out at temperatures above 100° F., this should be done at half the re-charge rate given above to avoid exceeding the maximum permissible electrolyte temperature of 120° F. If at the end of the charge the specific gravity varies by more than 5 points (*i.e.*, 0·005) from the figures given in Table "A", the gravity must be adjusted, either raised by the addition of electrolyte, the specific gravity of which at 60° F. would be 1·350, or lowered by the addition of distilled water. When adjusting the density, care must be taken not to leave too much electrolyte in the cell or cells ; any surplus over the top of the separators must be removed.

Testing Cells for Internal Short Circuit.—To test a cell suspected of being short-circuited, take the individual voltage of each cell of the battery, while it is on charge and also while not on charge, if possible also arrange to discharge the battery. If still fitted to the motor cycle this can be done by switching on the headlamp. In every case, the voltage of the bad cells will be much less than the others. If it is confirmed that the cell or cells are shorted internally, the battery must be replaced.

Special Precautions in Cold Climates

Where the temperature is frequently below 32° F. (freezing point), and not lower than − 21° F., the following special precautions should be taken :—

(1) *Limit of Discharge.*—The state of charge of the battery must not be allowed to fall below the condition indicated by the following specific gravities :—

1·200 specific gravity at			0° F.
1·245	,,	,,	−21° F.
1·265	,,	,,	−30° F.

otherwise the electrolyte may freeze and the battery be damaged through cracking of the cell containers.

Never allow a battery to stand in a discharged condition.

(2) *Electrolyte* in each cell should be level with the top of the separators. When topping up use distilled water. This should be done only during charging and preferably whilst the cells are gassing, so that the water becomes mixed with the electrolyte before it has time to freeze. Under these conditions the temperature of the topping up water is immaterial, but not more than a teaspoonful of water per cell must be added at a time.

LIGHTING AND ACCESSORIES

Headlamp

The headlamp incorporates a double filament main bulb and a pilot bulb for use when parking or when driving in well-lighted streets. The main bulb gives either a normal driving light or a dipped beam for use when meeting oncoming traffic, the changeover from the main to the dipped beam being effected by a switch located on the handlebars.

Removing Lamp Front and Reflector.—To remove the lamp front and reflector press back the fixing clip at the bottom of the lamp. When replacing the front, locate the top of the rim first, then press on at the bottom and secure by means of the fixing clip. To remove the bulb holder, press back the securing springs.

Setting and Focusing Headlamp.—The headlamp must be set so that the main driving light beam is parallel with the road surface. If necessary adjust the lamp by slackening the two fixing bolts, moving the lamp until the correct position is obtained and finally tightening the two bolts. To obtain the best driving light the bulb must be correctly focused in the reflector. To adjust the position of the bulb, remove the lamp front as described above. Slacken the clamping clip at the back of the reflector and move the bulb holder backwards and forwards until the best lighting is obtained. When the best position has been found, fully tighten the clamping screw.

Cleaning.—Care must be taken when handling the reflector to prevent it from becoming finger-marked. It can, however, be cleaned by polishing with a fine chamois leather. Metal polishes must not be used on the reflector. The black body of the lamp can be cleaned with any good car polish.

Replacement of Bulbs

When the replacement of a bulb is necessary it is important not only that the same size bulb is fitted, but that it has a high

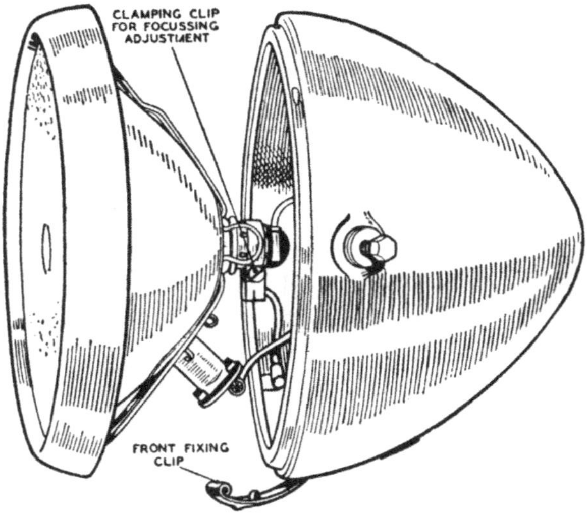

FIG. 89.—HEADLAMP WITH FRONT AND REFLECTOR REMOVED.

efficiency and will focus in the reflector. Provided the correct number genuine Lucas bulb is fitted as a replacement, these conditions will be fulfilled.

When fitting a main headlamp bulb care must be taken to fit it the correct way, *i.e.*, with the dipped beam filament above the centre filament. The correct replacement bulbs are as follows :

Headlamp	Main	Lucas No. 70	6 volt	24/24 watt
	Pilot	Lucas No. 200	6 volt	3 watt
Tail Lamp		Lucas No. 200	6 volt	3 watt

Cables

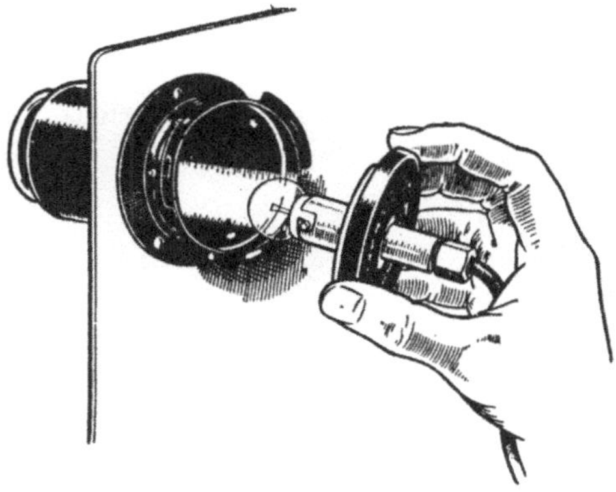

FIG. 90.—TAIL LAMP REMOVAL.

Before making any alterations to the wiring or removing the switch from the panel, disconnect the positive lead at the battery to avoid the danger of short circuits. The lead, about 1 ft. long, from the positive battery terminal is connected to the lead from the switch by means of a brass connector shown in Fig. 91. The connector is insulated by a rubber sleeve which must be pushed back to allow the connector to be unscrewed. Do not allow the brass connector to touch any metal part of the engine as this will short-circuit the battery. When connecting up again, pull the rubber sleeve over the connector.

Lighting Switch

The switch, together with the ammeter, is incorporated in a small panel. The panel can be removed when the fixing screw is withdrawn and the oil-pipe disconnected. The colour scheme and the diagram of connections are shown in the wiring diagram. When making connections to the switch, bare the end of the cable for about $\frac{3}{8}$ in., twist the wire strands together and turn back about $\frac{1}{8}$ in. so as to form a small ball. Remove the grub screw from the appropriate terminal

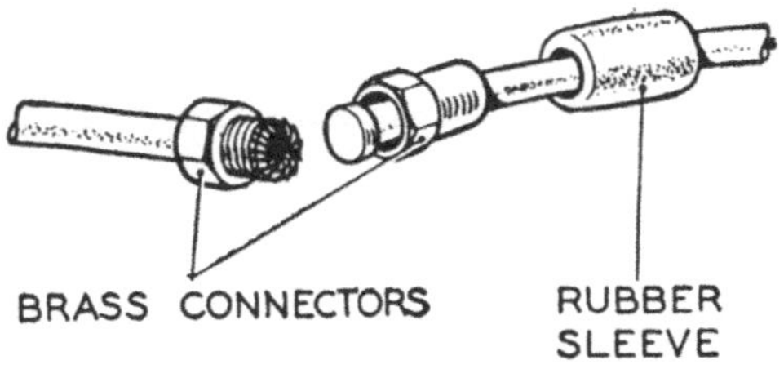

FIG. 91.—BATTERY CONNECTOR.

and insert the wire so that the ball fits in the terminal post. Now replace and tighten the grub screw; this will compress the ball to make a good electrical connection.

Horn

Electric horns are adjusted to give their best performance before leaving the Works and will give a long period of service without any attention. If the horn becomes uncertain in action or does not vibrate, it has not necessarily broken down. The trouble may be due to a discharged battery or a loose connection or short circuit in the wiring of the horn.

The performance of the horn may be upset by the fixing bolt working loose, or by the vibration of some part adjacent to the horn. To check this, remove the horn from its mounting, hold it firmly in the hand by its bracket and press the push. If the note is still unsatisfactory, the horn may require adjustment.

Method of Adjusting Horn

The adjustment of a horn does not alter the characteristics of the note, but takes up wear of vibrating parts which, if not corrected, results in loss of power or roughness of tone.

If the horn is used repeatedly when badly out of adjustment, due usually to unsuccessful attempts at adjustment, the horn may become damaged due to the excessive current which it will take. When testing, do not continue to operate the push if the horn does not sound. If, when the push is operated, the horn does not take any current (indicated by an ammeter connected in series with the horn) it is possible that the horn has been adjusted so that its contact breaker is permanently open.

After adjusting, note the current consumption, which must not exceed 4–5 amperes. A horn may give a good note, yet be out of adjustment and taking an excessive current.

When adjusting, do not attempt to unscrew the nut securing the tone disc or any other screw in the horn.

The adjustment is made by turning the adjustment screw, usually in a clockwise direction. The underside of the screw is serrated, and the screw must not be turned for more than two or three notches before re-testing. If the adjustment screw is turned too far in a clockwise direction, a point will occur at which the armature pulls in but does not separate the contacts.

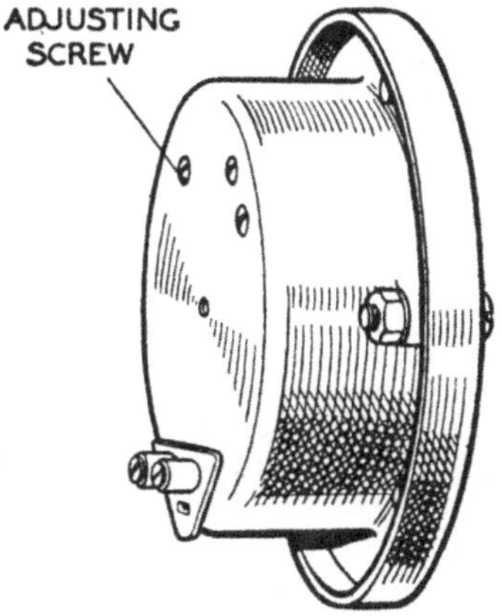

FIG. 92.—SHOWING HORN ADJUSTING SCREW.

Locating Faults in Lighting Circuit

(1) *If the headlamp does not light*, and the bulb proves to be in order: Test battery with voltmeter to ascertain that current is available. Then with switch in "Head" position check with voltmeter between terminal "2" on switch and a good earth on the frame of the motor cycle. If no reading, check between terminal "3" and earth, when reading will indicate a fault in the switch. If a reading is obtained between terminal "2" and the frame of the motor cycle, connect voltmeter between terminal "2" and lamp reflector. No reading will prove there is a poor earth, due either to a break in the short lead from the reflector to body of the lamp or between the lamp body and the frame of the motor cycle.

If a reading is shown the fault will be either in the cables to the headlamp (via the dipper switch) or a faulty dipper switch. To test, connect voltmeter to centre terminal of dipper switch and earth; a reading will prove wire is intact. To test the dipper switch, connect voltmeter to first one and then the other of the remaining terminals and earth (making sure that the dipper switch is in the correct position in each case). If no reading, dipper switch is faulty; if a reading is given carry on test with voltmeter from terminals at main bulb adaptor and earth, making sure to move over dipper switch for each test. Similar tests, as given in the first paragraph above, should be made to locate faults in the pilot light circuit except that terminal " 4 " should be used instead of " 2 ".

(2) *To locate a fault in the tail lamp circuit* switch on to " Head " position, when, if the headlamp lights, the battery and circuit to the switch will have been proved in order. Then connect voltmeter between terminal " 5 " and earth to test switch. If switch is satisfactory connect voltmeter between " 5 " and an earth on the tail lamp. No reading will prove poor earth between tail lamp and motor-cycle frame. To check cable from switch to tail lamp connect voltmeter from tail lamp terminal to earth : no reading will show fault in cable.

" MAGDYNO " EQUIPMENT

The Lucas electrical equipment dealt with above is for the latest twin-cylinder models. On the majority of Triumph motor cycles prior to these models, the electrical equipment incorporated a " Magdyno " (either single- or twin-cylinder type) in place of the separate dynamo and magneto. The " Magdyno " is a combined dynamo and magneto, the dynamo being mounted above the magneto and driven through gears from the magneto driving shaft.

Dynamo

The dynamo can be treated as a separate unit when it is removed from the " Magdyno ". To do this remove screw securing dynamo band clip (B_1, Fig. 93) and also nut (A_1, Fig. 93) securing the dynamo to the " Magdyno " body. The instructions for this dynamo are the same as described earlier in this chapter with the following exceptions :—

(1) *Dismantling.*—The armature is secured to the driving end bracket by means of an armature shaft nut. It will be necessary to remove this before removing armature from driving end bracket.

(2) The driving end bearing retaining plate is secured by two small screws and a longer screw known as the dynamo fixing screw. This screw locates in the casting of the " Magdyno " body.

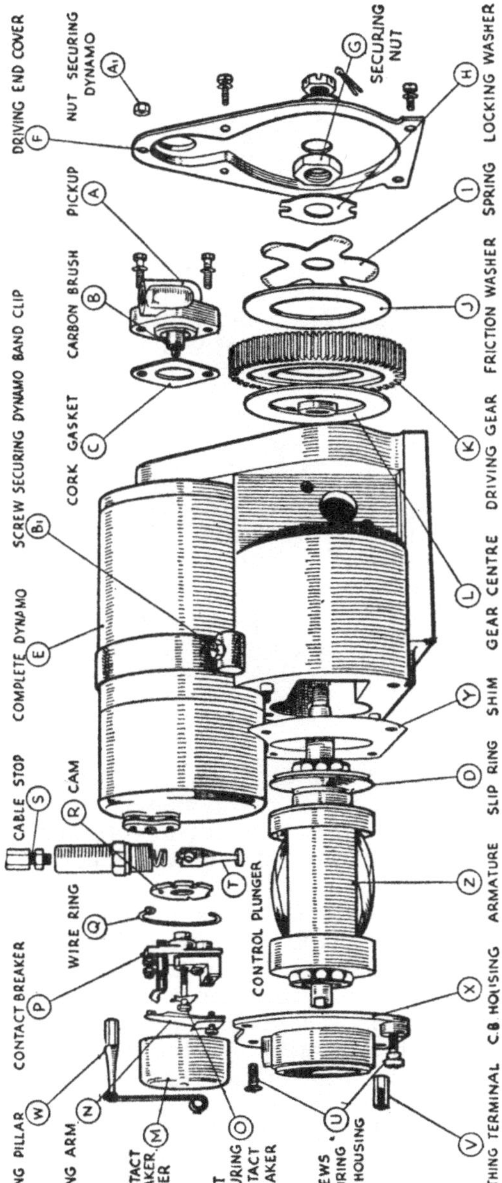

FIG. 93.—EXPLODED VIEW OF "MAGDYNO" (SINGLE-CYLINDER TYPE).

MAGNETO

Routine Maintenance of Magneto—Every 3,000 Miles

(1) *Lubrication* (*single-cylinder*).—The cam is lubricated by a wick, contained in the contact-breaker base, which must be given a few drops of thin machine oil. To get at the wick remove the spring arm (*A*, Fig. 94) carrying the moving contact and withdraw the screw (*B*) carrying the wick. At the same time, after unscrewing (*H*), remove the contact breaker and take from its housing (*C*) the tappet which operates the contact-

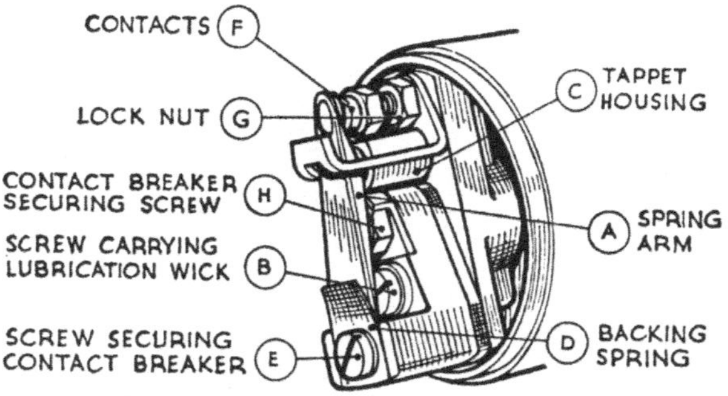

FIG. 94.—CONTACT BREAKER.

breaker spring and lightly smear with thin machine oil. When replacing the contact-breaker components, see that the small backing spring (*D*) is fitted immediately under the securing screw (*E*) and spring washer and that the bent portion faces outwards.

(2) *Lubrication* (*twin-cylinder*).—As described on page 166.

(3) *Adjustment* (*single-cylinder*).—After dismantling the contact breaker in order to lubricate, the contact setting should be checked. Turn the engine until the contacts (*F*, Fig. 94) are fully opened and insert the gauge provided, 0·010–0·012 in. thickness, between the contacts. If the setting is correct, the gauge should be a sliding fit. If there is an appreciable variation from the gauge, slacken the lock-nut (*G*) and turn the contact screw by its hexagon head, until the gap is set to the gauge. Tighten the lock-nut after making the adjustment.

(4) *Adjustment* (*twin-cylinder*).—As described on page 167.

Routine Maintenance Every 6,000 Miles (Magneto)

Contact Breaker Cleaning (single-cylinder).—Remove the contact-breaker cover and examine the contacts. If they are dirty, they must be cleaned by polishing with a very fine carborundum stone or very fine emery cloth; afterwards wipe away any dirt or metal dust with a petrol-moistened cloth. Cleaning of the contacts is made easier if the spring arm carrying the moving contact is removed as described on page 194. Examine the spring arm of the contact breaker and wipe away any rust.

Adjust as described on page 194.

Contact Breaker Cleaning (twin-cylinder).—As described on pages 167 and 168.

Testing Magneto in Position (Single- and Twin-Cylinder)

As already described on page 168, noting that there is, of course, only one ignition H.T. cable in the case of the single-cylinder unit.

Contact Breaker (single-cylinder).—Examine the contact breaker. If the contacts are badly burned they should be renewed by a replacement contact set. To remove the moving contact, withdraw the securing screw (*E*, Fig. 94) taking care not to lose the small backing spring (*D*). When refitting, this spring must be fitted immediately under the securing screw and spring washer with the bent portion facing upwards. To remove the fixed contact slacken the lock-nut and remove the contact by turning the hexagon head.

Contact Breaker (twin-cylinder). As described on page 168.

Renewing H.T. Cables (single- and twin-cylinder).—As described on page 168.

High-Tension Pick-ups and Slip-ring (single- and twin-cylinder).—As described on pages 168 and 170.

Dismantling Magneto

Take off the driving end cover (*F*, Fig. 93) by unscrewing the four countersunk head screws. To dismantle the slipping clutch it will be necessary to use a jig (Fig. 95) to hold the larger gear whilst the securing nut is being undone. This consists simply of a length of ¼ in. diameter mild steel rod bent to a flat U, the ends being cut short with their centres $3\frac{3}{16}$ in. apart, so that one can be slipped in the hole in the wheel whilst the other is engaged with the hole in the top of the casting through which the dynamo securing stud usually goes. The $\frac{7}{16}$-in. box spanner can then be used on the securing nut (*G*, Fig. 93) which unscrews in the normal left-hand direction. Note that the tab of the locking washer (*H*) must be bent back first.

Remove the locking washer (*H*), clutch spring (*I*), friction washer (*J*), driving gear (*K*) and gear centre (*L*).

Removing Magneto Armature.—On a single-cylinder type, take off the contact-breaker cover (*M*, Fig. 93), remove the spring

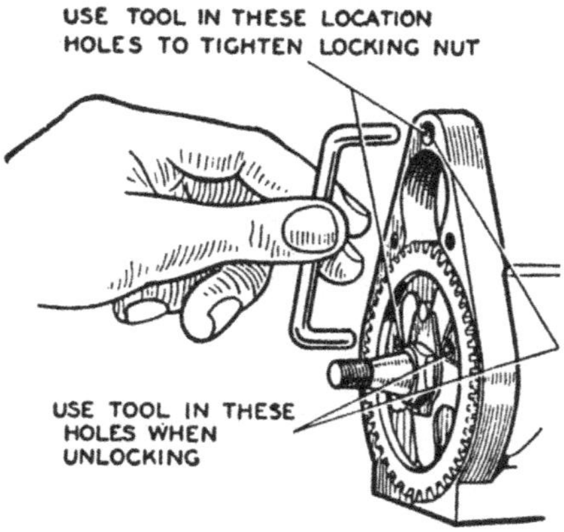

FIG. 95.—DISMANTLING SLIPPING CLUTCH.

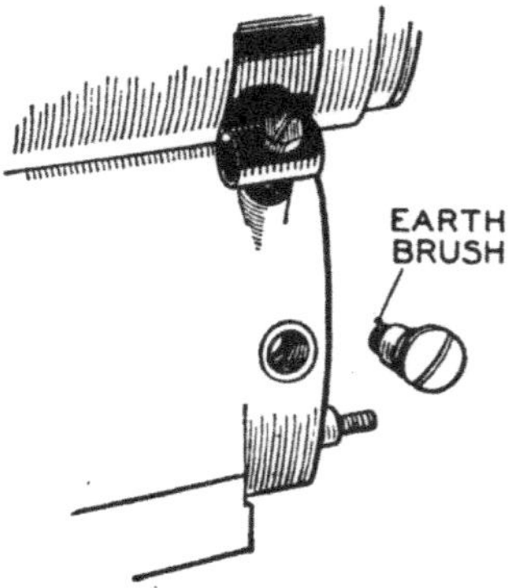

FIG. 96.—EARTH BRUSH LOCATION.

arm (*N*) carrying the contact, unscrew the bolt (*O*) securing the contact breaker (*P*), and draw the contact breaker off the shaft. Spring the wire ring (*Q*) securing the cam (*R*) out of its location in the contact-breaker housing, and remove the cam. Unscrew the cable stop (*S*) of the timing control and remove the control plunger (*T*).

On the twin-cylinder type remove the hexagon-headed screw and withdraw the contact breaker. Remove the cam which is a sliding fit, bend back tag and unscrew and withdraw the pillar which secures the timing control cable and plunger.

Remove the pick-up holder(s) and the small earthing brush (Fig. 96) which will be found on the side of the " Magdyno ". Unscrew the screws (*U*, Fig. 93), earthing terminal (*V*) and pillar (*W*) from the contact breaker end-plate (*X*) and remove the plate from the "Magdyno", together with the shims (*Y*). The armature (*Z*) can then be drawn out of the machine. There is no need to put a keeper across the magnet as it retains its magnetic properties more or less indefinitely. Although it loses a certain immaterial amount of power in the first removal of the armature, subsequent removals do not affect it. Do not allow the magneto body to come in close contact with any iron filings, as they may become attracted to the magnet and cause the armature to bind. The outer and inner journals can be removed as described on page 170.

Testing Magneto

If no test apparatus is available, a rough check of the armature windings can be made by means of a 2-volt battery and an ammeter, as follows :—

(1) Screw the contact-breaker retaining screw (*O*, Fig. 93) into the end of the armature shaft. Connect one terminal of the battery to the screw with the ammeter in series. Connect the other battery terminal to the metal body of the armature. The ammeter will then record the current taken by the primary winding—this should be approximately 4 amperes.

(2) To check the secondary winding of the armature, connect a piece of H.T. cable to the brass insert of the slip-ring and hold the other a little way from the armature core. If the lead from the battery which was connected to the core is then flashed quickly on and off the core, a spark should occur between the H.T. cable and the core. No spark at these points indicates a fault either in the armature windings or in the condenser and a replacement armature must be fitted.

Reassembling Magneto

Wash out bearings with white spirit, dry thoroughly and repack with high-melting-point grease. Fit ball races as already described on page 171. The serrated fibre washer must be fitted behind the journal to prevent electric current from damaging the bearing.

Place the armature in the body of the magneto and refit the contact-breaker end-plate, taking care that the end shims are in position and tighten the securing screws. Check the armature for end play. The armature should revolve freely when turned by hand but no end play should be felt. Adjust by adding or removing shims under the contact-breaker end plate until the adjustment is correct.

Single-Cylinder Type.—Fit the timing control plunger (*B*, Fig. 97) and spring and secure by tightening the cable stop (*C*). Locate the cam (*D*) in the contact-breaker housing with the timing control plunger in its correct slot as shown, and secure by springing the circlip into its location in the housing. (Note that the cam is fitted with its flat side towards the armature.)

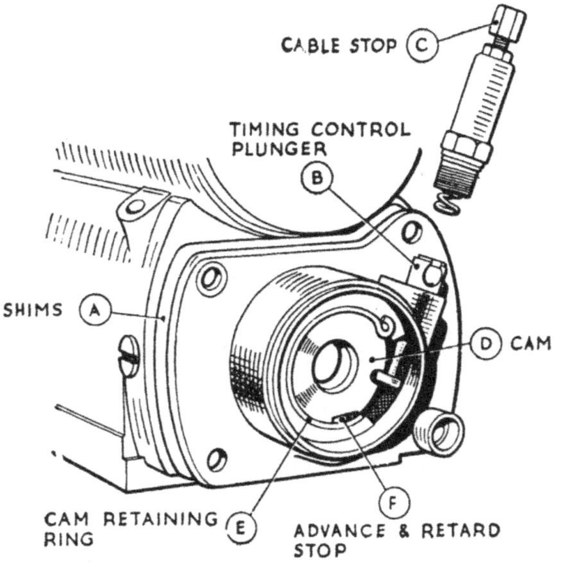

FIG. 97.—CONTACT-BREAKER HOUSING SHOWING ADVANCE AND RETARD MECHANISM.

Fit the contact-breaker body (*A*, Fig. 98) in position on the location at the end of the shaft after making sure that the tappet (*B*) is free and is located correctly in its guides. Place the contact-breaker securing screw (*C*) and locking plate (*D*) in position; tighten and lock by bending up the tags of the locking plate.

Refit the spring contact arm (*E*) with the backing spring (*F*) in its correct position. Check the contact-breaker gap and if necessary adjust to correct setting.

Twin-Cylinder Type.—Fit the timing control pillar, plunger and cable. Locate the cam in position so that the small slot in the cam locates into the control plunger.

Refit contact-breaker assembly after lubricating contact-breaker spring and rocker-arm pivot, taking care that the projecting key on the tapered shaft fits into the keyway cut in the armature shaft. Tighten the hexagon-headed screw with care. It must not be too slack, nor must undue force be used.

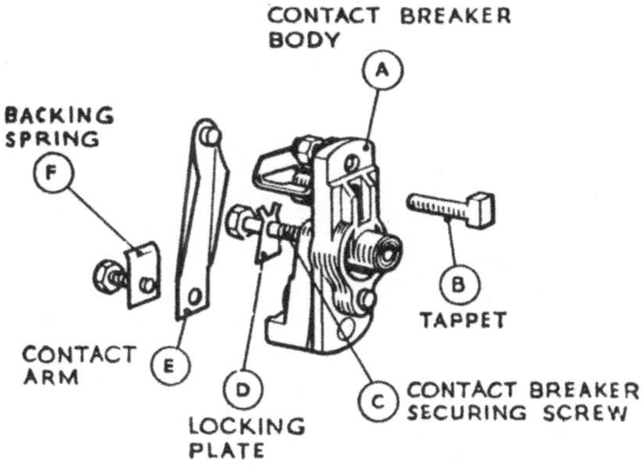

FIG. 98.—CONTACT-BREAKER ASSEMBLY.

Reassembling and Testing Slipping Clutch

Key the gear centre (*A*, Fig. 99) on to the spindle, replace the driving gear (*B*), friction washer (*C*), clutch spring (*D*), locking washer (*E*), and secure by tightening the fixing nut (*F*). The

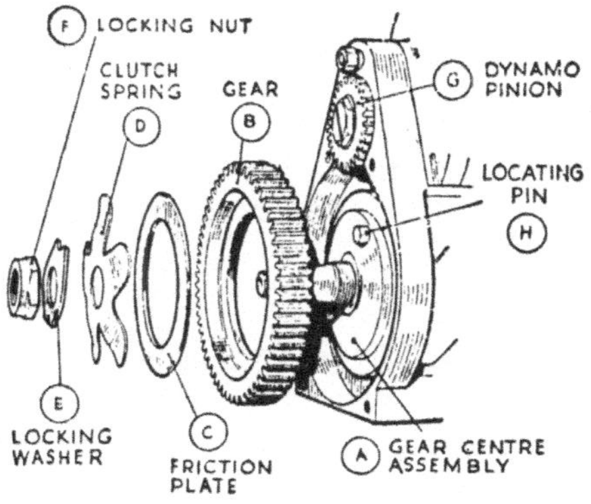

FIG. 99.—SLIPPING CLUTCH ASSEMBLY.

U-shaped jig must be used to prevent rotation of the shaft while tightening the nut.

After assembling, the setting of the clutch must be checked. This can easily be done by locking the driving gear and applying a steady load on the driving spindle as shown in Fig. 100. The clutch should slip with a load of 4–10 ft.-lb. or more, *i.e.*, a 4–10 lb. pull measured on a spring balance via a spanner 1 ft. long.

After setting, by slackening or tightening the securing nut (*F*, Fig. 99), prevent further movement by bending up the tab of the lock washer (*E*).

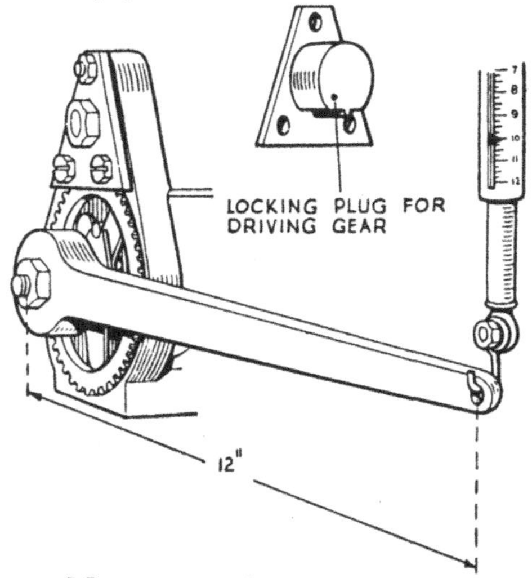

FIG. 100.—METHOD OF CHECKING CLUTCH SETTING.

Refit the dynamo and pack the gears with high-melting-point grease. Secure the drive end cover in position with the gasket correctly located.

Replace the pick-up, first checking that the brush moves freely and that the cork gasket is free from cracks.

Refit the earthing brush.

Lamps

Some machines were fitted with a headlamp which incorporated a panel containing ammeter and light switch, no panel being fitted on the tank.

Ignition Timing (Magdyno)

The ignition is correctly timed when, with the magneto fully advanced and the contact-breaker points just breaking, the piston is three-eighths of an inch from top dead centre.

Take off the contact-breaker cover and rotate the magneto until the contact points are just opening. Make certain that the magneto is rotated in the correct direction—anti-clockwise as viewed from the drive end.

Remove the sparking-plug and turn the engine by engaging a gear and, with the machine on the stand, rotating the rear wheel in the correct direction for forward travel. When both valves are closed, bring the piston up to top dead centre, inserting the graduated spoke through the sparking-plug hole to test the position. First making certain that the piston is on true top dead centre, rotate the engine backwards until the piston has dropped three-eighths of an inch. This amount can be measured on the graduated spoke without difficulty. Now check up to see that the magneto is correctly positioned—fully advanced and contact-breaker points just breaking—and, being careful that neither the engine nor the magneto moves, tighten up the nut on the armature shaft. Finally turn over the engine a few times and then make a check by again positioning the piston three-eighths of an inch before top dead centre and ascertaining that the contact-breaker position is correct. Though this operation calls for great care in order to make certain that the timing is exactly correct, the job presents no real difficulty.

	2H–2HC.	3S–3SE–3SW.	3H–3HW.	3T.	5S–5SE–5SW.	5H.
Engine.						
Bore, mm.	63	70	70	55	84	84
Stroke, mm.	80	89	89	73·4	89	89
Capacity, c.c.	249	343	343	349	493	493
Engine type.	O.H.V. Single	S.V. Single	O.H.V. Single	O.H.V. Twin	S.V. Single	O.H.V. Single
B.H.P. @ R.P.M.	13 @ 5200	12 @ 4800	17 @ 5200	19 @ 6300	15 @ 4800	23 @ 5000
Compression ratio.	6·92 : 1	5·3 : 1	6·7 : 1 *	6·3 : 1	5·6 : 1	6·1 : 1
Lubrication.	All dry-sump.					
Big-End(s) type.	Roller	Roller	Roller	Plain	Roller	Roller
Clearance (twin).	—	—	—	0·0015	—	—
Side-play.	0·003–0·015	0·003–0·015	0·003–0·015	0·0085–0·0090	0·003–0·015	0·003–0·015
Roller size.	¼ × ¼	¼ × ¼	¼ × ¼	—	¼ × ¼	¼ × 5/16
Big-end (external diameter).	0·9617–0·9619	0·9617–0·9619	0·9617–0·9619	1·3730–1·3735	0·9617–0·9619	1·2017–1·2020
Shafts and Tappets.						
Timing shaft (diameter).	0·9995–1·00	0·9995–1·00	0·9995–1·00	1·249–1·2485	0·9995–1·00	0·9995–1·00
	0·9995–1·00	0·9995–1·00	0·9995–1·00	—	0·9995–1·00	0·9995–1·00
Drive shaft (diameter).	1·1246–1·1248	1·1246–1·1248	1·1246–1·1248	0·9994–0·9998 0·8115–0·8125	1·1246–1·1248	1·1245–1·250
Camshaft bush (internal).	0·5610–0·5620	0·5610–0·5620	0·5610–0·5620	0·8740–0·8745	0·5610–0·5620	0·5610–0·5620
Idler gear bush (internal).	0·4995–0·5005	0·4995–0·5005	0·4995–0·5005	0·4995–0·5005	0·4995–0·5005	0·4995–0·5005
Tappet guide (internal).	0·437–0·438	0·437–0·438	0·437–0·438	0·3120–0·3125	0·437–0·438	0·437–0·438
Tappet (external).	0·4355–0·4365	0·4355–0·4365	0·4355–0·4365	0·3110–0·3115	0·4355–0·4365	0·4355–0·4365
Piston and Barrel.						
Bore diameter.	2·4796–2·4803	2·7555–2·756	2·7555–2·756	2·1649–2·1654	3·3066–3·3071	3·3066–3·3071
Maximum permissible wear.	0·006	0·006	0·006	0·004	0·006	0·006
Piston clearance :						
Top of skirt.	0·007–0·008	0·010–0·011	0·008–0·0085	0·0045–0·0050	0·010–0·011	0·009
Bottom of skirt.	0·005–0·006	0·007–0·008	0·004–0·0045	0·0030–0·0035	0·007–0·008	0·005
Ring gap.	All 0·010–0·014					
Ring clearance (comp.).	All 0·001–0·003					
(scraper).	All 0·0005–0·0025					
Valves.						
Valve guide diameter (internal).	0·3120–0·3130	0·3445–0·3450	0·3120–0·3130	0·3120–0·3130	0·3445–0·3450	In. 0·3432–0·3442 Ex. 0·374–0·375
Valve stem diameter (inlet).	0·3095–0·3100	0·3407 0·3412	0·3095–0·3100	0·3095–0·3100	0·3407–0·3412	0·3412–0·3407
Valve stem (diameter (exhaust).	0·3083–0·3088	0·3407–0·3412	0·3083–0·3088	0·3095–0·3100	0·3083–0·3088	0·372–0·3715
Valve seat angle.	All 45°					
Tappet clearance (cold).	0·001	In. 0·004 Ex. 0·007	0·001	0·001	In. 0·004 Ex. 0·007	0·001

* 6·3 with flat top piston.

† 1951 Model 32 @ 6500.

(All dimensions are in inches

DATA

	5T.	6S.	T70.	T80.	T90.	T100.
	63	84	63	70	84	63
	80	108	80	89	89	80
	498	597	249	343	493	498
	O.H.V.	S.V.	O.H.V.	O.H.V.	O.H.V.	O.H.V.
	Twin	Single	Single	Single	Single	Twin
	27 @ 6300	18 @ 4800	16 @ 5800	20 @ 5700	28–29 @ 5800	30 @ 6500 †
	7 : 1	5·6 : 1	7·7 : 1	7·5 : 1	7·08 : 1	7·8 : 1 ‡
	All dry-sump					
	Plain	Roller	Roller	Roller	Roller	Plain
	0·0015	—	—	—	—	0·0015
	0·030–0·0305	0·003–0·015	0·003–0·015	0·003–0·015	0·003–0·015	0·030–0·0305
	—	¼ × 5/16	¼ × ¼	¼ × ¼	¼ × 5/16	—
	1·4360–1·4365	1·2017–1·2020	0·9617–0·9619	0·9617–0·9619	1·2017–1·2020	1·4360–1·4365
	0·9994–0·9998	0·9995–1·00	0·9995–1·00	0·9995–1·00	0·9995–1·00	0·9994–0·9998
		0·9995–1·00	0·9995–1·00	0·9995–1·00	0·9995–1·00	
	1·1246–1·1248	1·1245–1·1250	1·1246–1·1248	1·1246–1·1248	1·1245–1·1250	1·1246–1·1248
	0·8115–0·8125					0·8115–0·8125
	0·8740–0·8745	0·5610–0·5620	0·5610–0·5620	0·5610–0·5620	0·5610–0·5620	0·8740–0·8745
	0·4995–0·5005	0·4995–0·5005	0·4995–0·5005	0·4995–0·5005	0·4995–0·5005	0·4995–0·5005
	0·3120–0·3125	0·437–0·438	0·437–0·438	0·437–0·438	0·437–0·438	0·3120–0·3125
	0·3110–0·3115	0·4355–0·4365	0·4355–0·4365	0·4355–0·4365	0·4355–0·4365	0·3110–0·3115
	2·4803–2·4796	3·3066–3·3071	2·4796–2·4803	2·7555–2·756	3·3066–3·3071	2·4803–2·4796
	0·006	0·008	0·006	0·006	0·006	0·006
	0·006–0·0065	0·010–0·011	0·007–0·008	0·008–0·0085	0·011	0·006–0·0065
	0·004–0·0045	0·007–0·008	0·005–0·006	0·004–0·0045	0·008	0·004–0·0045
	All 0·010–0·014					
	All 0·001–0·003					
	All 0·0005–0·0025					
	0·3120–0·3130	0·3445–0·3450	0·3120–0·3130	0·3120–0·3130	In. 0·3432–0·3442 Ex. 0·374–0·375	0·3120–0·3130
	0·3095–0·3100	0·3412–0·3407	0·3095–0·3100	0·3095–0·3100	0·3412–0·3407	0·3095–0·3100
	0·3090–0·3095	0·3412–0·3407	0·3083–0·3088	0·3083–0·3088	0·372–0·3715	0·3090–0·3095
	All 45°					
	0·001	In. 0·004 Ex. 0·007	0·001	0·001	0·001	0·001 §

‡ 1951 Model 7·6 : 1. § 1951 Model In. 0·002; Ex. 0·004.

unless otherwise indicated.)

	2H–2HC.	3S–3SE–3SW.	3H–3HW.	3T.	5S–5SE–5SW.	5H.
Valves (contd.)						
Rocker hole diameter (internal).	0·5620–0·5630	—	0·5620–0·5630	0·4995–0·5005	—	0·4995–0·5005
Rocker Spindle diameter.	0·5605–0·5610	—	0·5605–0·5610	0·4990–0·4995	—	0·4990–0·4995
Timing (Valve).						
Inlet opens B.T.D.C.	47½°	19°	47½°	22°	19°	26½°
Inlet closes A.B.D.C.	68½°	60°	68½°	66°	60°	62½°
Exhaust opens B.B.D.C.	69½°	67°	69½°	63°	67°	75½°
Exhaust closes A.T.D.C.	46½°	29°	46½°	25°	29°	20½°
Ignition.						
Timing fully advanced.	3/8	5/16	3/8	11/32 †	5/16	3/8
Contact point gap.	All 0·012					
Plug gap.	All 0·015–0·018					
Oil Pump.						
Body, scavenge diameter.	0·37475–0·37525	0·37475–0·37525	0·37475–0·37525 0·25025–0·25475	0·43725–0·43775	0·37475–0·37525	0·37475–0·37525
Body, feed diameter.	0·25025–0·25475 — —	0·31225–0·31275 — —	0·31225–0·31275 — —	0·31225–0·31275 0·37475 (1950–1) 0·37525 (1950–1)	0·31225–0·31275 — —	0·31225–0·31275 — —
Plunger, scavenge diameter.	0·3746–0·3749	0·3746–0·3749	0·3746–0·3749 0·2496–0·2500	0·4371–0·4374	0·3746–0·3749	0·4736–0·3749
Plunger, feed diameter.	0·2496–0·2500 — —	0·3121–0·3125 — —	0·3121–0·3125 — —	0·3121–0·3125 0·3746 (1950–1) 0·3749 (1950–1)	0·3121–0·3125 — —	0·3121 0·3125 — —
Carburetter.						
Bore.	7/8	15/16	1	7/8	15/16	1⅛
Main jet.	110	120	160	120	120	200
Slide.	5/4	5/4	6/4	5/4	5/4	29/3
Needle jet.	107	107	107	107	Std.	Std.
Needle.	5	6	6	5	6	29/29
Normal needle position.	3	3	3	3	3	3
Clutch and Gearbox.						
Lost motion (clutch rod).	All 1/32					
(clutch cable).	All 1/16					
Number of plates (steel).	4	4	4	4	4	5
Number of plates (corked).	3	3	3	3	3	4
Mainshaft (dia.):						
Bush end.	0·8098–0·8103	0·8098–0·8103	0·8098–0·8103	0·8098–0·8103	0·8098–0·8103	0·8098–0·8103
Bearing end.	0·7496–0·7498	0·7496–0·7498	0·7496–0·7498	0·7496–0·7498	0·7496–0·7498	0·7496–0·7498

† For low-octane fuel reduce to 9/32 in.

‡ 1951 Model 6 plates.

Data (*contd.*)

	5T.	6S.	T70.	T80.	T90.	T100.
	0·4995–0·5005	—	0·5620–0·5630	0·5620–0·5630	0·4995–0·5005	0·4995–0·5005
	0·4990–0·4995	—	0·5605–0·5610	0·5605–0·5610	0·4990–0·4995	0·4990–0·4995
	26½°	19°	47½°	47½°	26½°	26½°
	69½°	60°	68½°	68½°	62½°	69½°
	61½°	67°	69½°	69½°	75½°	61½°
	35½°	29°	46½°	46½°	20½°	35½°
	3/8	5/16	3/8	3/8	3/8	3/8
All 0·012						
All 0·015–0·018						
	0·43725–0·43775	0·37475–0·37525	0·37475–0·37525	0·37475–0·37525	0·37475–0·37525	0·43725–0·43775
	0·31225–0·31275	0·31225–0·31275	0·25025–0·25475	0·25025–0·25475	0·31225–0·31275	0·31225–0·31275
	0·37475 (1950–1)	—	—	—	—	0·37475 (1950–1)
	0·37525 (1950–1)	—	—	—	—	0·37525 (1950–1)
	0·4371–0·4374	0·3746–0·3749	0·3746–0·3749	0·3746–0·3749	0·3746–0·3749	0·4371–0·4374
	0·3121–0·3125	0·3121–0·3125	0·2496–0·2500	0·2496–0·2500	0·3121–0·3125	0·3121–0·3125
	0·3746 (1950–1)	—	—	—	—	0·3746 (1950–1)
	0·3749 (1950–1)	—	—	—	—	0·3749 (1950–1)
	15/16	1⅛	7/8	1	1⅛	1
	140	170	110	160	200	160 *
	6/3½	6/5	5/4	6/4	29/3	6/3½
	107	Std.	107	107	Std.	107
	6	6	5	6	29/29	6
	3	3	3	3	3	3
All 1/32						
All 1/16						
	5	5	4	4	5	5 ‡
	4	4	3	3	4	4 §
	0·8098–0·8103	0·8098–0·8103	0·8098–0·8103	0·8098–0·8103	0·8098–0·8103	0·8098–0·8103
	0·7496–0·7498	0·7496–0·7498	0·7496–0·7498	0·7496–0·7498	0·7496–0·7498	0·7496–0·7498

§ 1951 Model 5 plates.

* With air filter fitted use 150 main jet.

	2H-2HC.	3S-3SE-3SW.	3H-3HW.	3T.	5S-5SE-5SW.	5H.
Clutch and Gear-box (contd.)						
High gear bush layshaft (internal diameter).	0·7498-0·7505	0·7498-0·7505	0·7498-0·7505	0·7498-0·7505	0·7498-0·7505	0·7498-0·7505
High gear bush mainshaft (internal diameter).	0·8135-0·8145	0·8135-0·8145	0·8135-0·8145	0·8135-0·8145	0·8135-0·8145	0·8135-0·8145
Gear Ratios (*teeth*).						
Engine sprocket.	18	18	20 (3H) 19 (3HW)	19 19	20	23
(sidecar).	—	—	—	—	—	21
Clutch sprocket.	All 43					
Gearbox sprocket.	18	18	18	18	20	18
Rear chain sprocket (teeth).	All 46					
Chains.						
Primary—No. of pitches (solo).	74	74	75	74	75	78
Primary—No. of pitches (sidecar).	—	—	—	—	—	77
Size.	All $\frac{1}{2}$ × 0·335 × 0·305					
Rear—No. of pitches.	90	90	90	90	91	93
Size.	All $\frac{5}{8}$ × 0·400 × $\frac{3}{8}$					
Springs (*free length*).						
Fork (girder).	$8\frac{3}{16}$	$8\frac{3}{16}$	$8\frac{3}{16}$ (3H)	—	$8\frac{3}{16}$	$9\frac{5}{16}$
(sidecar).	—	—	—	—	—	$9\frac{1}{2}$
Telescopic fork.	—	—	—	$19\frac{5}{8}$	—	—
Valve spring (inner).	2	$1\frac{3}{4}$	2	$1\frac{1}{2}$	$1\frac{3}{4}$	$1\frac{7}{8}$
(outer).	$2\frac{1}{4}$	$2\frac{1}{2}$	$2\frac{1}{4}$	$1\frac{15}{16}$	$2\frac{1}{4}$	$2\frac{1}{16}$
Shock absorber.	$1\frac{11}{16}$	$1\frac{11}{16}$	$1\frac{11}{16}$	$1\frac{11}{16}$	$1\frac{11}{16}$	$1\frac{13}{16}$
Oil pump spring.	All $\frac{1}{2}$					
Release valve spring.	$\frac{1}{2}$	$\frac{1}{2}$	$\frac{1}{2}$	$1\frac{7}{32}$	$\frac{1}{2}$	$\frac{1}{2}$
Clutch pressure.	$1\frac{1}{2}$	$1\frac{1}{2}$	$1\frac{1}{2}$	$1\frac{1}{2}$	$1\frac{1}{2}$	$1\frac{1}{2}$
Gear box selector plunger spring.	All $2\frac{1}{2}$					
Gear box quadrant plunger spring.	All $1\frac{1}{2}$					
Tyres. Front.	3·25 × 19	3·25 × 19	3·25 × 19	3·25 × 19	3·25 × 19	3·25 × 19
Rear.	3·25 × 19	3·25 × 19	3·25 × 19	3·25 × 19	3·25 × 19	3·50 × 19
Petrol Capacity (gals.).	$3\frac{1}{4}$	$3\frac{1}{4}$	$3\frac{1}{4}$	$3\frac{1}{4}$	$3\frac{1}{4}$	$3\frac{1}{4}$
Oil Capacity (pints).	6	6	6	6	6	6
Wheelbase.	$52\frac{1}{2}$	$52\frac{1}{2}$	$52\frac{1}{2}$	$53\frac{1}{2}$	$52\frac{1}{2}$	54
Ground clearance.	All 6					
Saddle height.	$28\frac{1}{2}$	$28\frac{1}{2}$	$28\frac{1}{2}$	$28\frac{1}{2}$	$28\frac{1}{2}$	$29\frac{1}{2}$
Weight fully equipped (lbs.).	318	316	322	335	322	362
Overall length.	82	82	82	$82\frac{1}{2}$	82	84
Overall width.	All $28\frac{1}{2}$					
Brake diameter	All 7					
Brake diameter spring wheel.	—	—	—	8	—	—

* Heavier gauge for sidecar

† 3·00 × 20 (1938-40).

DATA (*contd.*)

5T.	6S.	T70.	T80.	T90.	T100.
0·7498– 0·7505	0·7498– 0·7505	0·7498– 0·7505	0·7498– 0·7505	0·7498– 0·7505	0·7498– 0·7505
0·8135– 0·8145	0·8135– 0·8145	0·8135– 0·8145	0·8135– 0·8145	0·8135– 0·8145	0·8135– 0·8145
22	23	18	20	23	22
19	21	—	—	21	19
All 43					
18	18	18	18	18	18
All 46					
78	78	74	75	78	78
77	77	—	—	—	77
All 1/2 × 0·335 × 0·305					
93	93	90	90	93	93
All 5/8 × 0·400 × 3/8					
9 5/16 9 1/2 20 * 1 5/8 2 1/32 1 13/16	9 5/16 9 1/2 — 2 2 1/4 1 13/16	8 3/16 — — 2 2 1/4 1 11/16	8 3/16 — — 2 2 1/4 1 11/16	9 5/16 9 1/2 — 1 7/8 2 1/16 1 13/16	9 5/16 9 1/2 20 * 1 5/8 2 1/32 1 13/16
All 1/2					
7/8 (1938) 3/4 (1939–40) 1 7/32 (1945–49)	1/2	1/2	1/2	1/2	3/4 (1939–40) 1 7/32 (1945–51)
1 1/2	1 1/2	1 1/2	1 1/2	1 1/2	1 1/2 (1938–50) 1·969 (1951)
All 2 1/2					
All 1 1/4					
3·25 × 19 † 3·50 × 19 4 ‡	3·25 × 19 3·50 × 19 3 1/4	3·00 × 20 3·25 × 19 3 1/4	3·00 × 20 3·25 × 19 3 1/4	3·00 × 20 3·50 × 19 3 1/4	3·25 × 19 † 3·50 × 19 4
6 ‖ 54	6 55	6 52 1/2	6 52 1/2	6 54	6 § 55
All 6					
29 1/2 361	29 1/2 356	28 1/2 316	28 1/2 320	29 1/2 362	29 1/2 361
84	84	82	82	84	84
All 28 1/2					
All 7					
8	—	—	—	—	8

‡ 3 1/4 (1938–9). § 8 (1938–47). ‖ 8 pints early post-war machines.

TECHNICAL DATA

	TR5.	6T.
Engine.		
Bore, mm.	63	71
Stroke, mm.	80	82
Capacity, c.c.	498	650
Engine type.	O.H.V. Twin	O.H.V. Twin
B.H.P. @ R.P.M.	25 @ 6000	34 @ 6300
Compression ratio.	6 : 1	7 : 1
Lubrication.	All dry-sump	
Big-End(s) type.	Plain	Plain
Clearance (twin).	0·0015	0·0015
Side-play.	0·030–0·0305	0·030–0·0305
Roller size.	—	—
Big-end (external diamter).	1·4360–1·4365	1·4360–1·4365
Shafts and Tappets.		
Timing shaft diameter).	0·9994–0·9998	0·9994–0·9998
Drive shaft (diameter).	1·1246–1·1248 0·8115–0·8125	1·1246–1·1248 0·8115–0·8125
Camshaft bush (internal).	0·8740–0·8745	0·8740–0·8745
Idler gear bush (internal).	0·4995–0·5005	0·4995–0·5005
Tappet guide (internal).	0·3120–0·3125	0·3120–0·3125

	TR5.	6T.
Shaftsand Tappets (contd.)		
Tappet (external).	0·3110–0·3115	0·3110–0·3115
Piston and Barrel.		
Bore diameter.	2·4803–2·4796	2·7953–2·7948
Maximum permissible wear.	0·006	0·006
Piston clearance :		
Top of skirt.	0·006–0·0065	0·006–0·0065
Bottom of skirt.	0·004–0·0045	0·004–0·0045
Ring gap.	All 0·010–0·014	
Ring clearance (comp.).	All 0·001–0·003	
(scraper).	All 0·0005–0·0025	
Valves.		
Valve guide diameter (internal).	0·3120–0·3130	0·3120–0·3130
Valve stem diameter (inlet).	0·0395–0·3100	0·3095–0·3100
Valve stem (diameter (exhaust).	0·3090–0·3095	0·3090–0·3095
Valve seat angle.	All 450	
Tappet clearance (cold).	In. 0·002 Ex. 0·004	0·001

	TR5.	6T.
Valves (contd.)		
Rocker hole diameter (internal).	0·4995–0·5005	0·4995–0·5005
Rocker Spindle diameter.	0·4990–0·4995	0·4990–0·4995
Timing (Valve).		
Inlet opens B.T.D.C.	26½°	26½°
Inlet closes A.B.D.C.	69½°	69½°
Exhaust opens B.B.D.C.	61½°	61½°
Exhaust closes A.T.D.C.	35½°	35½°
Ignition.		
Timing fully advanced.	⅜	⅜
Contact point gap.	All 0·012	
Plug gap.	All 0·015–0·018	
Oil Pump.		
Body, scavenge diameter.	0·43725–0·43775	0·43725–0·43775
Body, feed diameter.	0·31225–0·31275 0·37475–0·37525 (1950/1)	0·37475–0·37525
Plunger, scavenge diameter.	0·4371–0·4374	0·4371–0·4374

All Dimensions are in inches unless otherwise indicated.)

TECHNICAL DATA (*contd.*)

	TR5.	6T.
Oil Pump (contd.)		
Plunger, feed diameter.	0·3121–0·3125 0·03746 (1950/1) 0·3749	0·3746 0·3749
Carburetter.		
Bore.	1	1 †
Main jet.	160	190 *
Slide.	6/3½	6/3½
Needle jet.	107	107
Needle.	6	6
Normal needle position.	2	3
Clutch and Gear-box.		
Lost motion (clutch rod).	All 1/32	
(clutch cable).	All 1/16	
Number of plates (steel).	5	6
Number of plates (corked).	4	5
Mainshaft (diameter).	0·8098–0·8103	0·8098–0·8103
High gear bush layshaft (internal diameter).	0·7496–0·7498	0·7496–0·7498

	TR5.	6T.
Clutch and Gear-box (contd.)		
High gear bush mainshaft (internal diameter).	0·8135–0·8145	0·8135–0·8145
Gear Ratios (*teeth*).		
Engine sprocket.	21	24
(sidecar).	—	21
Clutch sprocket.	All 43	
Gearbox sprocket.	18	18
Rear chain sprocket (teeth).	46	46
Chains.		
Primary—No. of pitches (solo).	76	78
Primary—No. of pitches (sidecar).	—	80
Size.	½ × 0·335 × 0·305	
Rear—No. of pitches.	90	93
Size.	⅝ × 0·400 × ⅜	
Springs (*free length*).		
Fork (girder).	—	—
(sidecar).	—	—
Telescopic fork.	19¼	20 *
Valve spring (inner)	1⅝	1⅝
(outer).	2 1/32	2 1/32
Shock absorber.	1 13/16	1 13/16

	TR5.	6T.
Springs (*free length*) (contd.)		
Oil pump spring.	All ½	
Release valve spring.	—	—
Clutch pressure.	1 7/32 (1945–51) 1½	1 7/32 (1945–51) 1½ (1950) 1·969 (1951)
Gear box selector plunger spring.	All 2½	
Gear box quadrant plunger spring.	All 1¼	
Tyres. Front.	3·00 × 20	3·25 × 19
Rear.	4·00 × 19	3·50 × 19
Petrol Capacity (gals.).	2½	4
Oil Capacity (pints).	6	6
Wheelbase.	53	55
Ground clearance.	6½	6
Saddle height.	31	29½
Weight fully equipped (lbs.).	295	370
Overall length.	80	84
Overall (width).	29	28½
Brake diameter.	7	7
Brake diameter spring wheel.	8	8

(*All Dimensions are in inches unless otherwise indicated.*)

1950 Model with air cleaner fitted use 170 main jet. 1951 Model with air cleaner fitted use 140 main jet. † 1951 Model 1 1/16.

GEAR RATIOS FOR 1937 TO 1949 MODELS (Standard 18T Gearbox Sprocket)

	Standard Gears.				Wide Ratio Gears.				Close Ratio Gears.			
Engine Sprocket.	Top.	3rd.	2nd.	1st.	Top.	3rd.	2nd.	1st.	Top.	3rd.	2nd.	1st.
17 teeth	6·46	7·75	11·35	16·40	6·46	9·36	14·83	19·82	6·46	7·08	9·32	11·20
18 ,,	6·10	7·33	10·50	15·50	6·10	8·84	14·00	18·70	6·10	6·68	8·78	10·58
19 ,,	5·80	6·95	10·00	14·70	5·80	8·40	13·30	17·80	5·80	6·35	8·35	10·08
20 ,,	5·50	6·60	9·50	14·00	5·50	7·96	12·62	16·88	5·50	6·025	7·92	9·54
21 ,,	5·24	6·28	9·03	13·30	5·24	7·60	12·02	16·08	5·24	5·74	7·55	9·08
22 ,,	5·00	6·00	8·65	12·70	5·00	7·25	11·48	15·34	5·00	5·48	7·20	8·67
23 ,,	4·78	5·75	8·26	12·10	4·78	6·93	11·00	14·70	4·78	5·24	6·88	8·26
24 ,,	4·57	5·49	8·03	11·60	4·57	6·63	10·49	14·03	4·57	5·01	6·58	7·93
25 ,,	4·40	5·28	7·73	11·08	4·40	6·38	10·12	13·51	4·40	4·82	6·34	7·63
Gearbox Reduction	1·00	1·20	1·73	2·54	1·00	1·45	2·30	3·07	1·00	1·095	1·44	1·733

GEAR RATIOS FOR 1950 TO 1951 MODELS (Standard 18T Gearbox Sprocket)

	Standard Gears.				Wide Ratio Gears.				Close Ratio Gears.			
Engine Sprocket.	Top.	3rd.	2nd.	1st.	Top.	3rd.	2nd.	1st.	Top.	3rd.	2nd.	1st.
17	6·46	7·7	10·94	15·8	6·46	9·22	14·30	18·85	6·46	7·06	8·42	11·00
18	6·10	7·28	10·32	14·9	6·10	8·70	13·50	17·80	6·10	6·66	7·95	10·40
19	5·80	6·9	9·8	14·15	5·80	8·25	12·80	16·85	5·80	6·32	7·54	9·84
20	5·50	6·55	9·3	13·4	5·50	7·84	12·18	16·00	5·50	6·00	7·15	9·35
21	5·24	6·24	8·85	12·8	5·24	7·46	11·58	15·25	5·24	5·72	6·81	8·90
22	5·00	5·95	8·45	12·2	5·00	7·13	11·05	14·55	5·00	5·45	6·50	8·50
23	4·78	5·69	8·09	11·69	4·78	6·82	10·60	13·90	4·78	5·23	6·23	8·12
24	4·57	5·45	7·75	11·2	4·57	6·54	10·14	13·35	4·57	5·00	5·96	7·78
25	4·40	5·24	7·45	10·75	4·40	6·26	9·73	12·80	4·40	4·80	5·73	7·46
Gearbox Reduction	1·0	1·19	1·69	2·44	1·00	1·425	2·21	2·915	1·00	1·09	1·30	1·695

RECOMMENDED LUBRICANTS

	Castrol.	Essolube.	Price's Energol.	Shell.	Vacuum Mobiloil.	SAE.
Engine (single) Summer.	Grand Prix	Racer	60	Golden (Extra heavy)	D	60
Engine (twin) Summer.	XXL	50	40	Triple	BB	40
Engine (single) Winter.	Grand Prix	Racer	40	Triple	D	60
Engine (twin) Winter.	XL	40	30	Double	A	30
Gearbox.	XXL	50	40	Triple	BB	40
Primary Chain-case.	Castrolite	20	20	Single	Arctic	20
Front Fork (telescopic).	Castrolite	20	20	Single	Arctic	20
Spring Wheel Mechanism.	Castrolease Graphited Grease.	Esso Graphited Grease.	Belmoline CG	Retinax " RB "	Mobil Graphited Grease.	— —
Spring Wheel Bearings.	Castrolease Medium	Essogrease	Belmoline C	Retinax " RB "	Mobil Grease No. 2	—
Grease Gun.	Castrolease Medium	Essogrease	Belmoline C	Retinax " RB "	Mobil Grease No. 2	—
Easing Rusted Parts.	Castrol Penetrating oil	Esso Penetrating Oil	Price's Penetrating Oil	Shell Donax " P "	Mobil Spring Oil	—

A running-in compound (containing colloidal graphite) in the engine oil and an upper cylinder lubricant in the petrol are recommended when running-in new and reconditioned engines.

INDEX

INDEX

INDEX

Made in the USA
Middletown, DE
05 November 2026

36546288R10129